BIOLOGICAL MACROMOLECULES

A Series of Monographs

SERIES EDITORS

SERGE N. TIMASHEFF and GERALD D. FASMAN

Graduate Department of Biochemistry
Brandeis University
Waltham, Massachusetts

VOLUME 1

Poly-α-Amino Acids: Protein Models for Conformational Studies
edited by Gerald D. Fasman

VOLUME 2

Structure and Stability of Biological Macromolecules
edited by Serge N. Timasheff and Gerald D. Fasman

VOLUME 3

Biological Polyelectrolytes
edited by Arthur Veis

VOLUME 4

Fine Structure of Proteins and Nucleic Acids
edited by Gerald D. Fasman and Serge N. Timasheff

IN PREPARATION

Subunits in Biological Systems
edited by Serge N. Timasheff and Gerald D. Fasman

BIOLOGICAL POLYELECTROLYTES

BIOLOGICAL POLYELECTROLYTES

Edited by
Arthur Veis

Department of Biochemistry
Northwestern University Medical School
Chicago, Illinois

1970

MARCEL DEKKER, INC., New York

PREFACE

Living tissues are composed of complex mixtures of strongly interacting macromolecules at fairly high total solute concentrations. In the past several decades there has been a major emphasis in biochemical investigations on the detailed structure and chemistry of the biological macromolecules–proteins, nucleic acids and polysaccharides. The aim of much of this work has been to provide a basis for interpreting function in terms of specific chemical and topological properties of the macromolecules. However, control of the dynamic life processes, such as transport and metabolism, does not appear to depend entirely upon the chemical reactivities and structures of the individual macromolecules considered as independent moieties. A part of the control resides in the organization and structure of the cell and tissue components imposed by the requirements of the interactions of the biopolymers within the structural unit under consideration.

A detailed understanding of the biopolymer interactions important to the organization of living systems can be considered at two levels. One can approach the problem from the point of view of specific, and hence selective interactions, where local site contacts are of utmost importance. One may also choose to examine the interaction problem in more general terms where one considers forces of longer-range nature which depend primarily on molecular configuration and the distribution of electrical charges and polar groups over the molecular surface. An alternate way of viewing these factors might be to state that one can build up to the complex system by assembly of data concerning very specific systems, such as studies of the substrate specificity of an enzyme and the relation of that specificity to the constitution and configuration of the active site. One can build up understanding in a different way by asking what the general features of the behavior of polymers (and polyelectrolytes in particular) in solution contribute to the organization of mixtures of polymers. This latter approach is obviously more complicated and, at present, cannot yield results of such clear-cut and chemically satisfying nature as those, for example, relating the detailed arrangement of the active site of lysozyme or ribonuclease.

As more becomes known about the structures and properties of individual macromolecules, however, the chances for success in understanding tissue organization on broader thermodynamic and electrostatic principles will increase. A renewed interest in these aspects is evident in many current biochemical investigations.

This present brief volume is concerned with reviewing the groundwork basic to this more generalized approach to tissue architecture. The chapters have been assembled to cover the three main types of tissue biopolymers and to discuss these in terms of their polyelectrolyte character and overall molecular configuration. We have avoided, as much as possible, detailed consideration of conformation and conformational stability except as these are varied by interaction or ionic effects. In the final chapter, the concept of ordering of mixtures of biopolymers on the basis of entirely nonlocalized and nonspecific interactions is introduced.

It is recognized, of course, that in their development, the processes of selection in real living systems must have refined each system into what appears as the expression of very specific local interaction effects at all levels of organization but it is hoped that consideration from the more general approaches will add to our appreciation of the nature of the very complex ensembles of molecules that make up all living systems.

No attempt has been made to provide introductory or background chapters of more general nature. Four books, in particular, can provide most of the essential basic theory. Still unsurpassed as a prime reference in polymer chemistry and the thermodynamic treatments of polymer solutions is "Principles of Polymer Chemistry" (Cornell University Press, 1953) by P. J. Flory. The basic background for the understanding of the electrostatic theory will be found in the excellent monograph "Polyelectrolyte Solution: A Theoretical Introduction" (Academic Press, Molecular Biology Series, Vol. 2, 1961) by Stuart A. Rice and Mitsuru Nagasawa. A more general book dealing with biological polymers and polyelectrolytes at a more easily readable but rigorous level is "Macromolecules in Solution" (High Polymers, Vol. 21, Wiley, 1965) by Herbert Morawetz. Finally, the early work in this field, with particular regard to proteins and forming an excellent base for Chapter I is "Proteins, Amino Acids and Peptides" (Reinhold, 1943) by E. J. Cohn and J. T. Edsall.

Arthur Veis

Chicago, Illinois
1970

CONTRIBUTORS TO THIS VOLUME

Frederick A. Bettelheim, *Chemistry Department, Adelphi University, Garden City, New York*

David B. S. Millar, *Naval Medical Research Institute, Bethesda, Maryland*

Robert Steiner, *Naval Medical Research Institute, Bethesda, Maryland*

Serge N. Timasheff, *Graduate Department of Biochemistry, Brandeis University, and Pioneering Laboratory of Physical Biochemistry, Waltham, Massachusetts*

Arthur Veis, *Department of Biochemistry, Northwestern University Medical School, Chicago, Illinois*

CONTENTS

BIOLOGICAL POLYELECTROLYTES

CHAPTER 1

POLYELECTROLYTE PROPERTIES OF GLOBULAR PROTEINS

Serge N. Timasheff

GRADUATE DEPARTMENT OF BIOCHEMISTRY
BRANDEIS UNIVERSITY†

AND

PIONEERING LABORATORY OF PHYSICAL BIOCHEMISTRY
WALTHAM, MASSACHUSETTS

I. INTRODUCTION

Globular proteins are natural high polymers; they are characterized in general by the fact that their polymeric chains are folded in a unique way for each protein to form a highly compact, rigid, generally solvent-impenetrable structure which can be approximated normally by simple geometric models, such as a sphere or prolate or oblate ellipsoids of revo-

† This work was supported in part by NIH Grant No. GM-14603.

1

lution of axial ratios not very different from unity. The very particular electrolyte properties of globular proteins stem from the fact that they are actually copolymers of some twenty types of monomeric units, the amino acids, of which several can become positively or negatively charged depending on the medium in which they are immersed. Furthermore, in any given conformation of a globular protein, the many ionizable groups in it are present in definite spacial relationships with respect to each other. Changes in the nature of the medium in which they are immersed frequently induce changes in the overall configuration of a protein. Thus, the polyelectrolyte properties of a given protein may differ quite considerably from one medium to another. In this chapter a number of properties of globular proteins which stem from their polyelectrolyte nature will be discussed: these will include the electrostatic free energy of an isolated protein molecule and its relation to environment and molecular geometry, the interactions which globular proteins undergo in solution and their behavior in transport phenomena. First, however, it seems desirable to review briefly a few pertinent features of the structure of globular proteins.

II. SOME STRUCTURAL CONSIDERATIONS

A. Basic Structure of Globular Proteins

As stated above, globular proteins are natural copolymers of some twenty types of amino acids, of which eight possess chains capable of ionization. The differentiation between a globular and fibrous protein is essentially operational. Globular proteins are small. In general they have molecular weights in the range of 5000 to half a million. In the native state their molecules are compact and are characterized by low intrinsic viscosities, usually between 3.3 and 4 cc/g. This differentiates them from highly elongated rigid proteins such as collagen or myosin, or random flexible chain structures, which may have intrinsic viscosities anywhere from tens to thousands (in cubic centimeters/gram) (*1*). Furthermore, the intrinsic viscosity of a globular protein in the native state remains essentially constant over broad ranges of pH and ionic strength of the medium. Some proteins, such as γ-globulin, are found to exhibit viscosities which are significantly higher than those of the vast majority of globular proteins (for γ-globulin, $[\eta] = 7$ cc/g), but much too low to classify them among fibrous or random structures. It has been found that the structure of γ-globulin consists of several rigid compact regions joined by flexible joints (*2*); other proteins, such as serum albumin, which have normal viscosities for globular proteins in the zone of electric neutrality of the protein, undergo conformational transitions when they are highly charged with a concomitant large increase in viscosity. Both of these types of protein are normally regarded as globular and their properties will in general fall into the realm of the present chapter.

On a level of higher resolution, the compact structure is composed of a polymeric chain or several chains folded in a unique manner for any given protein. Thus, while the chain folding may be complicated, as is evident from the cases for which it has been determined so far [e.g., myoglobin (3), lysozyme (4), ribonuclease (5), chymotrypsin (6)], it is identical, as a time average, for all molecules of a given protein in a given conformation state. This folding which constitutes the secondary and tertiary structures of a protein is, in turn, a reflection of the order in which the monomeric units (amino acids) are arranged in the polymeric (polypeptide) chains. A number of amino acid sequences are known at present and their analysis indicates that this order is close to random, if the sequence of any protein is examined, but again it is identical for all the molecules of a given protein (once all the genetic and other variants have been absolutely purified), reflecting the genetic code in the DNA controlling its synthesis. In such a complicated copolymer, as a protein polypeptide chain, it might be expected naively that many different folded forms could occur under any given set of conditions. This is not true, however, as the secondary and tertiary structures of a protein reflect the tendency of any given sequence of amino acids to find a mutual arrangement of its residue in three-dimensional space which would result in a state of minimal free energy. The actual final three-dimensional structure may not correspond to an overall state of minimum free energy, due to kinetic effects (7). Thus, as the protein molecule comes off the ribosome, the portion synthesized first may assume a conformation of minimal free energy for it, forcing the rest of the chain as it comes off into the medium to assume a conformation of lowest free energy consistent with the pre-folded moiety of the molecule. Indeed, it is known that in lysozyme the amino terminal end of the chain is in the interior of the molecule (4), consistent with the fact that the synthesis starts from that end and proceeds to the carboxy-terminal group (8). Such an inherent instability in the protein molecule renders it highly susceptible to changes in its conformation with small changes in environment, as the tenuous balance between structure maintaining and structure disrupting factors is disturbed. Indeed, it may be estimated that the net free energy of structure stabilization of typical globular proteins is only of the order of 10–20 kcal/mole (7,9). Thus, small changes in the stabilization free energy in various parts of the molecule may result in local changes in conformation, which in turn may be transmitted to the rest of the structure and affect the polyelectrolyte properties of the entire molecule.

In a discussion of the polyelectrolyte properties of a globular protein, it would also seem necessary to examine the location of charges on such a molecule. From considerations of the forces which stabilize the three-dimensional structures of a protein, a basic rule has emerged that non-polar residues will be found in the interior of the molecule while polar residues will be on the outside; in aqueous medium, charged groups must be on the outside, except for the special case of ion pairs, since the burial of a charge in the nonpolar interior involves the expenditure of up to 100 kcal/mole of stabilization free energy (10). Furthermore, this factor is reinforced by the hydrophobic effect; namely, that it is entropically favorable to have the nonpolar (hydrocarbon) residues located in such a manner that they will be out of contact with water. Water in contact with nonpolar groups tends to form highly organized clathrate structures which are accompanied by a decrease in entropy. Thus, in order to keep the entropy of water at a maximum and the free energy of the system at a minimum, the nonpolar residues are forced to cluster together in the inner core of the globular structure, with the charged and most other polar groups present on the surface. Now, while such a folding causes a great loss of configurational entropy of the polypeptide chain, the last effect is overcome by the gain in water entropy. This structural feature has great significance for the electrolytic properties of globular proteins, since the general structural pattern becomes one of a core of low dielectric constant (it is estimated that the dielectric constant in the interior of a protein is 2–10) surrounded by a medium of high dielectric constant (for water, $D \sim 80$), with the charges located at the boundary between the two.

B. Charge Configuration of a Protein

The electric charge present on a protein molecule is the result of the ionization of a number of amino acids. Since these are weak electrolytes, their state of ionization is a strong function of the medium, i.e., of pH, ionic strength, dielectric constant, and temperature. The ionizable amino acids are listed in Table I together with typical values of the pK's enthalpies and entropies of ionization found in various proteins. The ionization of proteins has been discussed by various authors (11,14–17). At proper pH values, these groups ionize according to the reactions:

Acid Groups:

$$-G_aH \rightleftharpoons -G_a^- + H^+ \qquad K'_a = \frac{[H^+][G_a^-]}{[G_aH]}$$

Basic Groups:

$$-G_bH^+ \rightleftharpoons -G_b + H^+ \qquad K'_b = \frac{[G_b][H^+]}{[G_bH^+]}$$

$$(1)$$

TABLE I

Ionizable Groups in Globular Proteins[a]

Residue	Structure in acid form	$\left[\begin{array}{c} \mid \\ NH \\ \mid \\ H-C-R \\ \mid \\ C=O \\ \mid \end{array}\right]$ pK	$\Delta H,$ kcal/mole	$\Delta S,$ cal/deg mole
A. Negative (acid)				
Terminal (α) carboxyl	$-N-\overset{\underset{\mid}{H}}{\underset{\mid}{C}}-COOH$ H R	3.6–3.8	0–2	−12 to −20
Aspartic acid[b] R = —CH$_2$COOH Glutamic acid[b] R = —CH$_2$—CH$_2$COOH		4.0–4.8	0–2	−12 to −20
Tyrosine	R = —CH$_2$—⟨benzene⟩—OH	9.5–10.5	6–8	−20 to −25
Cysteine	R = —CH$_2$—SH	9.5		
Phosphate	R = R′—O—$\overset{\underset{\mid}{OH}}{\underset{\mid}{\underset{O}{P}}}$—OH	$pK_1 \sim 1$ $pK_2 \sim 6$		
B. Positive (basic)				
Histidine	R = —CH$_2$—C=CH $^+$HN NH C H	6.3–7.8	7–8	−8
Lysine	R = —(CH$_2$)$_4$—NH$_3^+$	9.6–10.4	11–14	−6 to −8
Arginine[c]	R = (CH$_2$)$_3$—NH—C$\overset{NH_2}{\underset{NH_2^+}{\diagdown}}$	>13	12–13	
Terminal (α) amino	—C—CH—NH$_3^+$ O R	7.5–8.5	11–14	−6 to −8

[a] Data complied from Tanford (11).

[b] Nozaki and Tanford (12) have determined that the pK's of aspartic and glutamic acid are 4.08 and 4.50, respectively.

[c] Nozaki and Tanford (13) report a pK of 13.74 for guanidine in 6 M solution.

The result is that either positive or negative charges are introduced into the protein molecule. [K_a', K_b' are the apparent dissociation constants for the acid (a) and basic (b) groups under the existing conditions; as we shall see below, they are related to, but not equal to, the intrinsic dissociation constants.]   Thus, if only proton binding can affect the charge on a protein, the total number of charges on a protein molecule, Z_t, at any given conditions, is given by:

$$Z_t = \sum_a G_a^- + \sum_b G_b H^+ = \sum_a \frac{K_a'[G_a H]}{[H^+]} + \sum_b \frac{[G_b][H^+]}{K_b'} \tag{2}$$

where the summation is carried out over all types of acid and basic groups.   The net charge at any instant, Z_p, is obtained by multiplying each group in Eq. 2 by its valence; thus

$$Z_p = \sum_b G_b H^+ - \sum_a G_a^- \tag{3}$$

Comparison of Eqs. 2 and 3 show that it is only at extreme pH's where all the groups are either protonated (net maximal positive charge) or dissociated (net maximal negative charge) that the absolute values of Z_t and Z_p will be equal.   At intermediate pH's, these two quantities will differ greatly.   Thus, at the pH at which the number of positive and negative charges on the protein is identical, the protein has a neutral net charge, i.e., $Z_p = 0$; Z_t, on the other hand, may be quite large.   For β-lactoglobulin, for example, $Z_p = 0$ at a pH at which $Z_t \to 80$, i.e., the majority of the 40 positively and 60 negatively charged groups are ionized. The net charge on a protein will be zero at a pH value, defined by:

$$[H^+]_{i.e.} = \sqrt{\frac{\sum K_a'[G_a H]}{\sum \frac{[G_b]}{K_b'}}} \tag{4}$$

In the absence of binding of other charged species to the protein, this pH will correspond to the isoelectric point, $[H^+]_{i.e.}$ As will be discussed below, the net charge on a protein molecule is in constant fluctuation; operationally we usually deal with the net average charge, $\bar{Z}_p$, i.e., the time average of Z_p.   By definition a protein is isoelectric at such conditions that $\bar{Z}_p$ is zero.   Ions other than hydrogen can also be bound by a protein making an additional contribution to the charge.   As a result, the net average charge on a protein due to the binding of all ions (A) is:

$$\bar{Z}_{p,\text{total}} = \bar{Z}_{p,\text{protons}} + \bar{Z}_{p,\text{other ions}} \tag{5}$$

where $\bar{Z}_{p,\text{other ions}}$ is given by

$$\bar{Z}_{p,\text{other ions}} = \sum_j k_j n_j z_j [A]_j \tag{6}$$

where k_j is the binding constant of ionic species j, z_j is its valence, n_j

is the number of binding sites for ion j on the protein molecule, and $[A]_j$ is the concentration of free ions of type j. Operationally, the isoelectric point is determined most easily by electrophoresis experiments: the pH at which the protein is found to be immobile in an electric field is said to be the isoelectric point under the given conditions. As can be seen from Eq. 4, the isoelectric point is independent of protein concentration, being a function solely of the binding of protons (or other ions) to various sites on the protein. Nevertheless, in the determination of the isoelectric point by electrophoresis, extrapolation to zero protein concentration is desirable because of hydrodynamic and other factors.

A related term is the isoionic point of a protein; this is defined as the state of the protein in solution such that only hydrogen and hydroxyl ions are present in addition to the polyelectrolyte. In such a case, electroneutrality gives for the isoionic pH, $[H^+]_{i.i.}$

$$[H^+]_{i.i.} = [OH^-] - \bar{Z}_p m_p \tag{7}$$

where m_p is the molar concentration of protein. Since the product $[H^+][OH^-] = K_w \sim 10^{-14}$ is a constant, it is evident that at a pH other than 7.0 an isoionic protein will not be also isoelectric, i.e., it will have a nonzero $\bar{Z}_p$. Furthermore, the charge on an isoionic protein will be a function of protein concentration, as will the pH of the solution, since electroneutrality will cause the hydrogen ion concentration to re-adjust itself in order to maintain the ionization equilibria of Eq. 1. In Fig. 1, the

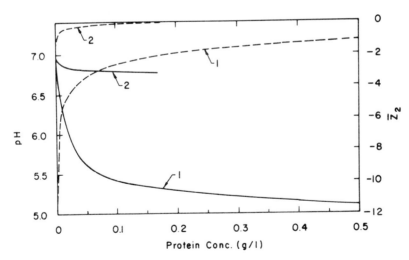

Fig. 1. Variation of the protein charge, $\bar{Z}_2$ (dashed lines) and pH (solid lines) of the aqueous solution of isoionic proteins as a function of protein concentration. (1) Bovine serum albumin; (2) conalbumin.

variations of pH and $\bar{Z}_p$ with protein concentration are shown for bovine serum albumin and for conalbumin which have isoelectric points of 5.3 and 6.75, respectively. It is quite striking that deviation of the isoelectric point from 7.0 by as little as one quarter of a pH unit can result in the presence of a significant charge on an isoionic protein at low concentration. As we shall see later, this property of an isoionic protein may make a considerable contribution to its activity coefficient.

Up to this point, nothing has been said either about the distribution of the charges on a protein molecule or of their interactions. A globular protein molecule being a close-to-rigid structure, the charged (or ionizable) groups on it are located in definite positions, i.e., in a given conformation their space coordinates are essentially fixed with respect to each other. Thus, as a first approximation, one may regard a protein molecule as a large cavity of low dielectric constant, on the surface of which are located points which may, under proper conditions, assume a positive or a negative charge. The locations of these points are identical from one molecule of the protein to another, and their distribution is determined by the three-dimensional folding of the polypeptide chains. As we shall see below, while it is convenient to regard this distribution as uniform over this surface, in actuality the probability of this being so is essentially negligible. The charge configurations of any given protein most probably involve areas with different densities of any given type of residue, while the distances between the individual point charges may also vary considerably. One of the consequences is that there may be rather complex patterns of interaction between the individual groups which cause the values of the individual apparent dissociation constants, K', to vary within broad limits and which impart various degrees of structural stability or destabilization to the molecule as a whole. These effects will be discussed in a later section. At this point it seems desirable to discuss one general type of interaction between the charged groups on a protein, one which is related to the K' values of the groups and can be discussed without considering the exact distribution of charges and their mutual electrostatic interactions. This interaction is related to the fluctuations in the structural and charge configurations which are constantly taking place on polyelectrolyte molecules.

C. Fluctuations

As stated above, charges on a protein are generated by reactions such as described in Eq. 1. Except in highly acidic solutions, the number of basic sites generally exceeds the number of protons bound to the molecule, so that the various groups, whether located on a single molecule or on different protein molecules, will be in a state of dynamic equilibrium with each other and with the solvent.   At any instant there will be present in the solution protein molecules which differ in the number and configuration of the protons bound to the molecules ($14,18,19$).   In general these differ little in free energy, and fluctuations will occur between them.   By taking the charge configuration which has the minimum free energy as the standard state, the probability, P_i, of any given state i of the molecule is related to the free energy of that state, ΔF_i^e, by

$$\frac{P_i}{P_0} = \exp\left(-\frac{\Delta F_i^e}{RT}\right) \tag{8}$$

where P_o is the probability of the standard state, R is the gas constant, and T is the thermodynamic temperature.   As a result, since the values of ΔF_i^e will be small for small variations in charge configuration, individual molecules in the solution will carry different electric charges at any given instant.   In an isoelectric protein, while the net average charge, $\bar{Z}_p$, is zero, molecules with charges of $0, +1, +2, \ldots, -1, -2, \ldots$ will be present.   This is exemplified in Fig. 2 for hemoglobin at pH 6.4 (14).   We see that while, as a time average, the protein is essentially neutral, at any instant a significant number of molecules have net charges between $+4$ and -4.   Thus, while the mean charge of the protein is zero, its mean square

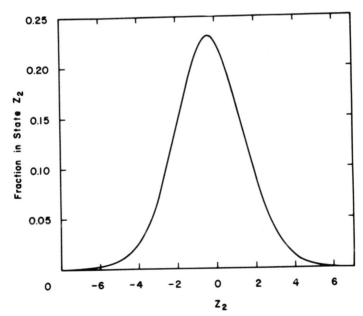

Fig. 2.   Distribution of forms with net charge $\bar{Z}_2$ for isoelectric hemoglobin (14).

charge, $\overline{Z^2}$, is not zero.   This phenomenon has important consequences for the electrostatic properties of proteins since, as will be shown in Section IV.B.2, it may result in a large effect on the activity coefficient of the protein (20,21) and actually lead to strong intermolecular interactions, as well as to make a contribution to the dipole moment of the protein (19). The magnitude of the charge fluctuation can be measured directly from the titration curve of the protein (18,22,23), as well as from light-scattering experiments (20,24,25).   It can be calculated using the equation of Kirkwood and Shumaker (20),

$$\overline{Z_{\mathrm{H}}^2} = \sum_j \frac{1}{2 + K_i'/[\mathrm{H^+}] + [\mathrm{H^+}]/K_i'} \tag{9}$$

where K_i' is the apparent dissociation constant of basic group i.   Since fluctuations will occur between any types of ions bound to the protein, the total mean square charge will be the sum of contributions from all such fluctuations (21):

$$\overline{Z^2} = \overline{Z_{\mathrm{H}}^2} + \sum_j \overline{Z_j^2} \tag{10}$$

where

$$\overline{Z_j^2} = \sum_j \frac{1}{2 + [A_j]k_j + ([A_j]k_j)^{-1}} \tag{11}$$

One final structural consideration of significance to the electrolyte properties of globular proteins should be mentioned.   While we have discussed globular proteins in terms of rigid structures with point charges located at discrete positions on their surfaces, this situation is far from realistic and in any rigorous discussion must be modified to include fluctuations in these positions, i.e., fluctuations in protein conformation.   Just as in the case of the charge configuration, the three-dimensional space coordinates of any given residue in the protein (and in particular on its surface) will be subject to fluctuations (26,27) due to the fact that the differences between the configurational free energies between different related protein conformations are small.   Indeed, hydrogen–deuterium exchange studies have shown that all the groups of a protein must come in contact with solvent over periods of less than one week, while most groups come in contact with solvent with a frequency of the order of once every hour or less (7,28). An ionizable group, then, will fluctuate between conformations which will cluster about the one which has the lowest free energy.   This may be treated in a similar manner to charge fluctuations.   Equation 8 shows that thermal fluctuations alone will permit the group in question to assume conformations which differ in free energy by 0.5 kcal/mole from the most favored one.   As a result, since the amino acid residues which carry charged side chains are most subject to such fluctuations because of their being on the surface of the molecule, it becomes impossible to assign an exact position in space to the charges relative to the center of mass of the protein molecule and, consequently, to all the other charges on it.   It is, therefore, impossible to make calculations of the electrostatic properties of proteins based on an exact assignment of positions of the charged residues. Such calculations may, however, be undertaken by using as a most probable configuration the one corresponding to the minimal free energy of the potential well that corresponds to the protein conformation under given conditions.   As a practical matter, at present, the best approximation to this conformation seems to be that deduced from X-ray crystallographic studies.   A fully rigorous calculation would require the calculation of probabilities of an extremely large number of related conformations and the averaging over all such conformations.   At present, this is a task beyond our possibilities.   As a time average, the free energy of a protein molecule will be somewhat higher than the minimal value, since fluctuations both in charge and conformation will result in small positive free energy increments.   The total probability of any given state j is small; it is given by Eq. 8, where now the free energy change is a sum of the free energy differences between the lowest energy state and the state in the instantaneous number and configuration of charges, as well as conformation of residues.   With the model of a globular protein established in terms of its charge and configuration and their fluctuations, we will proceed now to a discussion of several properties of these molecules which result from their polyelectrolytic nature.   The following topics will be discussed in order: (1) the electrostatic free energy of a globular protein, (2) its interactions with hydrogen ions, (3) its intermolecular interactions and (4) its behavior in transport phenomena.   No attempt will be made to discuss the related subject of internal electrostatic interactions, such as intramolecular ion pairs and dipolar interactions between ordered sequences of polypeptide chains inside a protein molecule (29,30).

III. ELECTROSTATIC FREE ENERGY OF GLOBULAR PROTEINS

A. The Linderstrøm-Lang Model

In examining the conformation or interactions of a protein molecule, it is frequently necessary to know its electrostatic free energy and the effects upon it of various changes in environment.

At constant temperature and pressure, a change in free energy is equal to the work done on a system; it follows, therefore, that the electrostatic free energy, F^e, of a protein is the work, W_{el}, done in increasing its charge from zero to $Z\epsilon$, i.e.:

$$F^e = W_{el} = \int_0^{Z\epsilon} \psi d(Z\epsilon) \tag{12}$$

where Z is the number of charges, ϵ is the protonic charge in esu, and ψ is the potential at its surface at any time during the charging process. Assuming that the protein molecule may be described as an impenetrable sphere, immersed in a continuous medium of dielectric constant, D, with the charges smeared out evenly over its surface, Linderstrøm-Lang (18) has shown that, with the applicability of the Debye–Hückel approximation, ψ and F^e are given by:

$$\psi = \frac{Z\epsilon}{D}\left[\frac{1}{b} - \frac{\kappa}{1 + \kappa a}\right]$$
$$F^e = \frac{Z^2\epsilon^2}{2D}\left[\frac{1}{b} - \frac{\kappa}{1 + \kappa a}\right] = Z^2 kTw \tag{13}$$

where b is the radius of the protein sphere, a is the distance of closest approach between the protein ion and small ions in the medium, κ is the well-known Debye–Hückel parameter (the reciprocal of the thickness of the double layer):

$$\kappa^2 = \frac{4\pi\epsilon^2}{DkT}\sum_i n_i z_i^2 \tag{14}$$

where k is Boltzmann's constant, T is the thermodynamic temperature, n_i is the molar concentration of ionic species i, and z_i is its valence. w is the work function in the acid–base equilibrium of proteins defined by Eq. 13. Its full significance will be discussed later.

While this model for a protein, shown on Fig 3a, departs strongly from the actual structure and charge distribution of real molecules, it has been found to give quite satisfactory results. One contributing factor to this is actually the fluctuating nature of the charge configuration of a protein; thus, as a time average, the charge configuration of a protein molecule would tend to approach the uniform smeared distribution of the model.

In calculating F^e it is necessary to know the radius, b, of the sphere equivalent to the actual molecule. Since the completely folded protein will include some solvent tightly bound to it, either as individual molecules on the inside or strongly bound ones at polar sites as well as molecules trapped in molecular crevices, this solvent must be included in this calculation. According to Tanford (31):

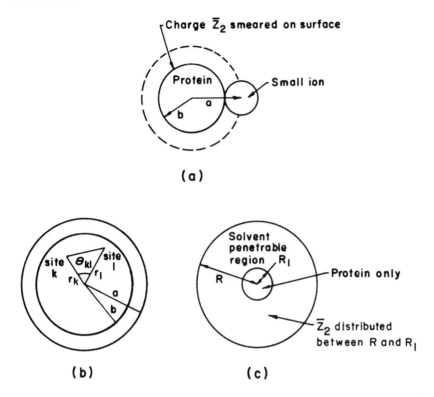

Fig. 3. Models for calculation of F^e. (a) Smeared charge model (18); (b) discrete charge model (10); (c) solvent permeable model (50).

$$b = \left[\frac{3}{4\pi} \frac{M}{N} (\bar{V}_2 + \delta_0 v_0^0) \right]^{1/3} \tag{15}$$

where M is the molecular weight, N is Avogadro's number, $\bar{V}_2$ is the partial specific volume of the protein, v_0^0 is the specific volume of the solvent usually assumed to be 1.00 for water, and δ_0 is the amount of bound solvent in grams per gram of dry protein, usually assumed to vary between 0.2 and 0.4 for globular proteins in an aqueous medium.

Alternately, the radius may be calculated from experimental data; for example, from hydrodynamic measurements, such as the sedimentation coefficient, $s_{20,w}^0$, since (32)

$$b = \frac{M(1 - \bar{V}_2 \rho_0)}{6\pi\eta N s_{20,w}^0} \tag{16}$$

where ρ_0 is the density of the solvent and η is the solvent viscosity; b may also be calculated from the radius of gyration, R_g, determined in a small angle X-ray scattering experiment since, if a spherical model is assumed for the protein,

$$b^2 = \tfrac{5}{3}R_g^2 \tag{17}$$

Typical values of b, obtained from sedimentation (Eq. 16) (*33–35*) and small angle X-ray scattering (Eq. 17) measurements (*36*), for β-lacto-globulin A in three states of molecular aggregation are compared in Table II with values derived from the corresponding electrostatic free energies

TABLE II

Values of b in Å for Equivalent Sphere Models of
β-Lactoglobulin A in Various States of Association

Species	Method of measurement		
	$s^0_{20,w}$	R_G	F^e
Monomer	20.2	18.0	
Dimer	27.0	27.7	27.0
Octamer	43.3	44.4	43.8

(Eq. 13), measured in titration experiments (*37*) (see below). This par-ticular protein seems to be a good case for these calculations, since its gross hydrodynamic structure is known and the actual deviations of the spherical model from the true shape are specified; thus, the monomer is close to spherical, the dimer is an aggregate of two slightly impinging spheres (*36,38*), while the octamer is a cubic array of eight spheres in 422 symmetry (*36,39*). Approximation of the three structures by spheres is seen to give good agreement between the effective radii, b, derived from the various experiments.

B. Departures from the Linderstrøm-Lang Model

1. Debye–Hückel Approximation

Equation 13 for the electrostatic free energy has been derived within the Debye–Hückel approximation, i.e., $\epsilon\psi/kT \ll 1$. Thus, its appli-cability is limited to the region in which $2|Z|w > 1$; since, as will be shown later, the parameter w varies between the limits of 0.2 and 0.002, depend-ing on the size of the protein and the ionic strength of the medium, this approximation will be valid in a region in which $|Z| < 5$–50; in most cases this limit will lie between $|Z| = 10$–15. To treat the case where $\epsilon\psi/kT > 1$, Nagasawa and Holtzer (*40*) have obtained a numerical solution of the

Poisson–Boltzman equation, keeping the assumption of a uniform smeared charge.† Their expression for the electrostatic free energy is:

$$F^e = \frac{Z\epsilon^2}{2kT} \psi_a + \frac{Z^2\epsilon^2}{2DkT} \frac{(a-b)}{(ab)} \tag{18}$$

where ψ_a is the computed electrostatic potential at the radius of closest approach of salt ions to the macro ion. Using the last term as an adjustable parameter, Nagasawa and Holtzer have fitted the titration curves of several protein, in the region of high charge, and have obtained values of $(a-b)$ between 0 and 2.5 Å; usually this value is assumed to be 2 Å. However, since in reality the protons on ionizable groups in a protein are not located on a spherical surface of radius b, the results of Nagasawa and Holtzer may be regarded as quite satisfactory.

2. Nonrandom Discrete Charge Model

The case of the discrete location of charges on the surface of a protein molecule has been examined by several authors (*10,42–45*). The most complete treatment is that of Tanford and Kirkwood (*10,42,43*) which was carried out within the Debye–Hückel approximation. The model selected was that shown in Fig. 3b. The molecule is regarded as an impenetrable sphere of radius b, with small ions capable of approaching to within a distance a from the center. The charges are located at p fixed positions, each located at the end of a vector r_k ($k = 1, 2, \ldots, p$), drawn from the center of the sphere; the distribution is no longer random and the positions are not restricted to the surface of the protein. The internal dielectric constant is designated by D_i. The charge at each point is $\zeta_k\epsilon$, where ζ_k is $+1$ for a positive charge and -1 for a negative charge. The final expression obtained for this model is:

$$F^e = W_{el} = \frac{\epsilon^2}{2b} \sum_{\substack{k=1\\k\neq l}}^{p} \sum_{l=1}^{p} \zeta_k\zeta_l(A_{kl} - B_{kl}) - \frac{\epsilon^2}{2a} \sum_{k=1}^{p} \sum_{l=1}^{p} \zeta_k\zeta_l C_{kl} \tag{19}$$

where

† It should be noted that, while numerical solutions of the Poisson-Boltzmann equation remove the limitations of the Debye-Hückel linear approximation, they introduce a new source of uncertainty; namely, the violation of the principle of superposition of fields (*41*). This is the result of the nonlinear form of such solutions, while field additivity requires linearity. Thus, while such solutions seem to result in better agreement with experiment than the Debye-Hückel approximation at higher charge and ionic strength, they can be regarded, at present, as no better than semi-empirical relations with a qualitative rationale of their significance.

$$A_{kl} = \frac{b}{D_i r_{kl}}$$

$$B_{kl} = \frac{1 - 2\delta}{D_i(1 - 2\rho_{kl}\cos\theta_{kl} + \rho_{kl}^2)^{1/2}} + \frac{1}{D\rho_{kl}}$$

$$\ln\left[\frac{(1 - 2\rho_{kl}\cos\theta_{kl} + \rho_{kl}^2)^{1/2} + \rho_{kl} - \cos\theta_{kl}}{1 - \cos\theta_{kl}}\right] \quad (20)$$

with

$$\delta = \frac{D_i}{D} \quad \text{and} \quad \rho_{kl} = \frac{r_k r_l}{b^2}$$

C_{kl} is a complicated function of the ionic strength, given by Tanford and Kirkwood (10). The factors involving A_{kl} and B_{kl}, respectively, represent the work of charging the sites in an unbounded medium of dielectric constant D_i, and the effect of restricting this medium to a cavity surrounded by a medium of high dielectric constant D, such as water. C_{kl} represents the interaction of the fixed sites with the small ions of the solvent. It vanishes at zero ionic strength, and at very low ionic strength its contribution to F^e reduces to $-Z^2\epsilon^2\kappa/2D(1 + \kappa a)$ since

$$\sum_k \sum_l \zeta_k \zeta_l = \left(\sum_k \zeta_k\right)\left(\sum_l \zeta_l\right) = Z^2 \quad (21)$$

and Eq. 20 becomes

$$F^e = W_{el} = \frac{\epsilon^2}{2b}\sum_{\substack{k=1 \\ k \neq l}}^{p}\sum_{l=1}^{p}\zeta_k\zeta_l(A_{kl} - B_{kl}) - \frac{Z^2\epsilon^2}{2D}\frac{\kappa}{1 + \kappa a} \quad (22)$$

The last term is identical with the ionic strength form of Eq. 13, so that the ionic strength dependence reduces to that of the smeared charge model. Tanford and Kirkwood have calculated that the approximation for C_{kl} of Eq. 22 is valid up to a value of $\kappa a = 0.5$; this would correspond to ionic strengths between 0.001 and 0.04 for typical globular proteins which have equivalent radii in the range of 20–60 Å.

Calculations of the electrostatic free energy of various models by Eqs. 19 and 22 have led Tanford (42) to the conclusion that the charges must be located close to 1.0 Å below the surface of the molecule. This is a reflection of the model which requires an abrupt jump at the surface from D_i to D. Tanford (43) has pointed out that in actuality, water molecules in contact with a charged site must be highly immobilized due to the charge-dipole interaction; as a result, the dielectric constant of the solvent at the charge must be lower than that in the bulk solution and the jump in D must occur at some distance from the charge. On the other hand, the charge cannot be buried deep within the low dielectric constant interior of the protein molecule, since the self-energy of such a charge increases very sharply with increasing depth in a cavity of low dielectric constant. Thus, burial of a charge would destabilize a protein molecule by 10–100 kcal/mole of free energy, i.e., an amount that could not be tolerated without inducing a major change in the conformation; the net free energy of structure stabilization in a globular protein is of the order of 10–20 kcal/mole.

A further complication, alluded to before, results from the fluctuations of protons between the fixed sites and the solvent (14,18,22), leading to fluctuations of charge and charge configuration.   If it is desired, therefore, to compare the calculated F^e with an experimental value, it becomes necessary to calculate the values for all combinations of charge and charge configuration and to obtain the properly weighted average, the probability of any such state being given by Eq. 8.   Since different configurations involve different sites at which charges are located, Tanford and Kirkwood (10) have carried out the averaging calculation over all possible configurations at a given total charge by setting p in Eqs. 19 and 22 equal to the number of sites rather than the number of charges.   The final step requires a knowledge of the distribution of the states with different net charges and an averaging over these.   In the case of the smeared charge approximation, this becomes rather simple, since the averaging over charge distributions in each state of charge becomes unnecessary.   Then, if at any instant the system contains N_i molecules of charge Z_i, the average electrostatic free energy, F^e, becomes:

$$\overline{F^e} = \frac{\epsilon^2}{2D}\left(\frac{1}{b} - \frac{\kappa}{1+\kappa a}\right)\frac{\Sigma N_i Z_i^2}{\Sigma N_i} = \frac{\overline{Z^2}\epsilon^2}{2D}\left(\frac{1}{b} - \frac{\kappa}{1+\kappa a}\right) \quad (23)$$

Within the smeared charge approximation, this effect may be significant only at pH values close to the isoelectric point, where the average charge of a protein is low and $\overline{Z^2} - \bar{Z}^2$ will be larger than $\bar{Z}^2$.   For example, in the case of conalbumin, when $\bar{Z} = 0$, $(\overline{Z^2})^{1/2} = 1.6$ (23); then, at an ionic strength of 1×10^{-3}, the mean electrostatic free energy calculated with Eq. 23 will be 0.2 kcal/mole.   Under conditions where the net average charge is somewhat removed from zero ($|Z| > \sim 5$), the difference between $\overline{Z^2}$ and $\bar{Z}^2$ becomes insignificant since fluctuation will result in species with charges both lower and higher than $\bar{Z}$ and the results of Eq. 23 become essentially indistinguishable from those of Eq. 13.

The effect of charge distribution on F^e has been calculated for simple models by Tanford (42).   For a cubic distribution of two positive and four negative charges, shown by models (a) and (b) of Fig. 4, if the corners are located 0.8 Å below the surface of a sphere 8 Å in radius,

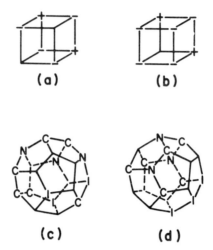

Fig. 4. Models for the distribution of charges at discrete positions on a sphere. (a) and (b): Two possible configurations of locating four negative and two positive charges on the apeces of a cube; (c) and (d): models for calculating F^e of Fig. 5. C, I, and N stand for carboxyl, imidazole, and amino groups, respectively. The dodecehedra are inscribed in spheres of radius 10 Å, with charges 1 Å below the surface. The distance between nearest neighboring vertices is 6.5 Å. Taken from (46).

setting $D_i = 2$ results in F^e of -2.5 and $+0.7$ kcal/mole for charge configurations (a) and (b), respectively, with $\bar{Z} = -2$. Thus, in structure (a), the attraction between the two positive charges and their three closely neighboring negative groups (along the edges of the cube) outweighs the mutual repulsion of the four negative charges along the diagonals. In structure (b), the repulsive forces outweigh the attractions with a net destabilization of the molecule. By applying Eq. 8, it is seen that, at any instant, configuration (a) is more probable than configuration (b).

Tanford (42) has also calculated F^e as a function of net average charge for two molecules of identical size and amino acid composition but different charge distribution, as shown in Figs. 4c and d, and compared the results with calculations using the smeared charge model. These results are presented in Fig. 5 for a salt-free system. Comparison with the results of Eq. 13 brings out the very striking result that, while the smeared charge model can never result in a net stabilizing electrostatic free energy, discrete distribution can result in a high degree of stabilization: -11 kcal/mole in the case of structure (d) when $\bar{Z} = -2$, while Eq. 13 predicts a destabilizing contribution of 0.84 kcal/mole. Tanford has pointed out, however, that the smeared charge model can be made to give negative values of F^e: Eq. 13 includes the self-energies of all the charges, i.e., the work of creat-

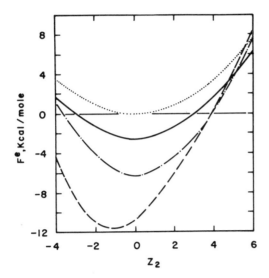

Fig. 5. F^e as a function of Z_2 for models (c) and (d) of Fig. 4. (· · ·) Eq. 13; (——) Eq. 24; (— · —) Eq. 19, model (c); (— —) Eq. 19, model (d).

ing the p individual charges, while Eq. 19 and 22 do not. Within the Debye–Hückel approximation, the self-energy of a charge is given by Eq. 13 with $Z = 1$ and $\kappa = 0$. Thus, their subtraction results in:

$$F^e = \frac{Z^2 \epsilon^2}{2D} \left(\frac{1}{b} - \frac{\kappa}{1 + \kappa a} \right) - (n_+ + n_-) \frac{\epsilon^2}{2Db} \qquad (24)$$

where n_+ and n_- are the number of positive and negative charges on the macroion, respectively. The corresponding values of F^e for structures (c) and (d) of Fig. 4 in the smeared charge approximations are also presented in Fig. 5. Quite evidently, accounting for the self-energies of the charges results in a structure stabilizing electrostatic free energy in the neighborhood of the isoelectric point. In most calculations, however, it is the change in F^e, i.e., ΔF^e which is of interest; therefore, since in most cases the discrete location of charges on a protein is not known and the self-energies remain unchanged during the process in question, it becomes most expedient to use Eq. 13 to calculate $\Delta F^e = F^e_{(1)} - F^e_{(2)}$.

3. Departure from the Solvent Impenetrable Spherical Model

While all the previous discussion was carried out in terms of a spherical model for proteins, it is well known that most globular proteins depart significantly from this model. Thus, the values of F^e calculated by Eqs. 13, 19, 22, and 23 still deviate from the correct F^e because of the assumption of spherical symmetry. Hill has derived an expression for ellipsoids of revolution which is valid in the absence of added electrolyte (47). For the case of finite ionic strength, Linderstrøm-Lang (48) has carried out numerical calculations of F^e for ellipsoidal ions with charges located along the major axis of the ellipsoid; this result, however, is of little interest for a discussion of globular proteins. In general, it can be stated that F^e for ellipsoids of moderate axial ratios (<5), a situation true for most proteins, will be of similar magnitude and somewhat smaller than values calculated using Eq. 13. This is due principally to the larger surface and smaller charge density of an ellipsoid than of a sphere of equal volume.

Other pertinent models that have been examined within the Debye–Hückel approximation are cylinders, which may serve as models for rigid proteins such as myosin, and spheres penetrable to solvent which correspond to partly denatured proteins, such as serum albumin at low pH or denatured proteins in which the —S—S bonds have remained intact. These will be examined in order.

For a cylinder of length L and radius b, Hill (49) calculated

$$F^e = \frac{Z'^2\epsilon^2 L}{D}\left[\frac{K_0(\kappa a)}{\kappa a K_1(\kappa a)} + \ln\frac{a}{b}\right] \tag{26}$$

where a is the distance of closest approach of a small ion to the axis of the cylinder, Z' is the number of smeared charges per unit of length of the cylinder, and $K_0(x)$ and $K_1(x)$ are modified Bessel functions of the second kind.

The case of a protein penetrable to solvent has been treated by Tanford (50). The model, shown in Fig. 3c, consists of a central impenetrable core of radius R_1, surrounded by a uniformly penetrable region out to radius R; the fraction of the volume $\frac{4}{3}\pi$ $(R^3 - R_1^3)$ accessible to small ions is designated by α^2. Then, if the dielectric constant of the solvent penetrable region has the bulk value D,

$$F^e = W_{el} = \frac{9Z^2\epsilon^2}{2\kappa^2 D(R^3 - R_0^3)}$$

$$\left\{\frac{1}{3} + \frac{1 + \kappa R}{\kappa^3(R^3 - R_0^3)}\frac{(1 + \alpha\kappa R)\dfrac{1 + \alpha\kappa R_1}{1 - \alpha\kappa R_1}e^{2\alpha\kappa(R - R_1)} - (1 + \alpha\kappa R)}{(1 + \alpha)\dfrac{1 + \alpha\kappa R_1}{1 - \alpha\kappa R_1}e^{2\alpha\kappa(R - R_1)} - (1 - \alpha)}\right\} \tag{27}$$

where R_0 is the radius of the sphere that would be made of the protein substance of the molecule; for a denatured or expanded protein, R_0 can be taken from the volume of the native protein molecule. Equation 27 has been used successfully to explain the experimental value of F^e of bovine serum albumin at conditions at which the molecule is expanded. It is interesting that Luzzati et al. have concluded from small angle X-ray scattering measurements that, under such conditions, this molecule can be described as a hard core surrounded by a loose solvent penetrable region (51).

In Eq. 27 it is assumed that the dielectric constant within the penetrable region of the protein is similar to that of the bulk solvent, limiting the applicability of this equation to highly diffuse structures. The intermediate case between this and a compact sphere can be treated by assigning a dielectric constant $D' = \alpha^2 D$ to the volume contained between R and R_1. In the case where the entire structure is penetrable, $R_1 = 0$ and Eq. 27 can be simplified accordingly.

The final case, which might occur if the protein contained a number of crevices penetrable to solvent but not to small ions and if the charges were distributed evenly and randomly along these crevices, can be described by a model examined by Hill (49). Here the charge is distributed uniformly throughout a sphere of radius b and dielectric constant D_i. Then, if all the other symbols have their previous meaning,

$$F^e = \frac{Z^2 \epsilon^2}{2Db} \left(1 - \frac{\kappa b}{1 + \kappa a} + \frac{D}{5D_i} \right) \tag{28}$$

Examination of the above equations leads to the conclusion that, at present, the most useful expression for F^e is still the one based on the smeared charge spherical model in a Debye–Hückel atmosphere (Eq. 13). It gives reasonably good agreement with experimental observations. The more complicated expressions are all applicable to special situations such as the one in which the charge distribution is completely known. This information is available so far for only a few proteins. As more structures become known, these expressions should become more and more applicable. The other equations are all specialized for particular organizations of protein matter in space and are, thus, of value only for some particular cases. Finally, it should be stated that F^e can be measured experimentally as a function of pH either from the titration curves (see below) of a protein or from moving boundary electrophoresis. Katchalsky et al. (52) have determined values of ψ measured from electrophoretic mobility ($\psi = 300u\eta C/D$), when η is the viscosity of the solvent, u is the electrophoretic mobility, and C is a factor depending on the size and shape of the particle, given by Henry and Gorin (see Eq. 93 below). They have found that values of F^e calculated from these values of ψ by Eq. 12 are in good agreement with values measured from titration data. This approach, however, is valid only within the limit that the thickness of the double layer is greater than the radius of the sphere, i.e., $\kappa b < 1$; for most proteins this condition is fulfilled at ionic strengths less than 0.01. Finally, nothing has been said in this section about the electrostatic free energy of fully denatured proteins, i.e., proteins with their covalent crosslinks (S—S bonds) broken and unfolded in a high concentration of urea or guanidine. These molecules are essentially flexible polymers and should be treated in terms of the theory of random coils (46).

4. Changes in F^e Due to Changes in Conformation and State of Aggregation

Frequently the processes of interest in proteins are accompanied either by changes in the conformation or in the molecular weight of the kinetic units in solutions, or both. For a process involving only a change in molecular charge, the quantity $\Delta F^e = F_1^e - F_2^e$ requires essentially only the application of Eq. 13 or of one of the more exact or special equations with the appropriate changes in $\bar{Z}$. When changes in conformation or molecular size or weight are involved, F_1^e and F_2^e must be properly calculated for each terminal case by taking into account the stoichiometry of the reaction as well as change in molecular weight, radius, charge and, if the exact equations (Eqs. 19 or 22) are used, charge distribution. Thus, when a change in the disperse state occurs, i.e., for the reaction $n\text{M} \rightleftharpoons \text{P}$, the electrostatic component, ΔF^e, of the free energy of aggregation is

$$\Delta F_A^e = \bar{F}_P^e - n\bar{F}_M^e \tag{29}$$

where M and P are the monomer and aggregate, respectively, and n is the degree of aggregation.

Since in most cases no detailed information is available on charge distribution, the present discussion will be restricted to the smeared charge spherical model (Eq. 13). For this case, Tanford and Epstein (53) give the equation.

$$\Delta F_A^e = \frac{n\bar{Z}^2\epsilon^2}{2D} \left(\frac{n^{2/3} - 1}{b} - \frac{n\kappa}{1 + n^{1/3}\kappa a} + \frac{\kappa}{1 + \kappa a} \right) \tag{30}$$

which is based on the assumption that the aggregate can be described as a sphere with a radius of $n^{1/3}b$, and the average net charge of $n\bar{Z}$. In many cases this is the best approximation that can be made in view of the lack of specific information on the geometry and charge of either the monomer or the aggregate. In the case where both are known, the appropriate F^e values can be calculated directly and subtracted from each other according to Eq. 29.

At present many enzymes and other proteins are known to be aggregates of small compact subunits; these may be reasonably expected to be close to spherical in shape. The calculation may be refined somewhat by evaluating the electrostatic free energy of the dimer intermediate, using the more exact Verwey-Overbeek potential (54) for the electrostatic interaction of two spheres ΔF^e_{dim}, with a center-to-center distance R:

$$\Delta F^e_{dim} = \psi_0^2 \frac{Db^2}{R} e^{-\kappa(R-2b)}\gamma \qquad (31)$$

ψ_0 is the surface potential of the spheres and γ is a complicated function of the ionic strength and particle separation which has been tabulated by Verwey and Overbeek. For two touching spheres (formation of the dimer), $R = 2b$ and the equation is simplified accordingly. A calculation of this type has been carried out for the β-lactoglobulin association from dimer to octamer. The shapes of the monomer, dimer, and octamer forms of this protein are known from X-ray crystallographic (38) and small angle X-ray scattering (36,39) studies (see page 14). Furthermore, from the titration curves the mean net charges, $\bar{Z}_i$, of the dimer and octamer are known as a function of pH (37). $\bar{Z}$ of the monomer was assumed to be half the value of the dimer at any pH. The geometries of the monomer and octamer are close to spherical, and their electrostatic

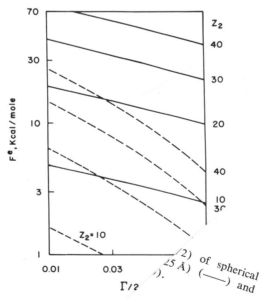

Fig. 6. Electrostatic free energy at var'
protein molecules in the compact imp'
after expansion to t'

free energies can be reasonably described by Eq. 13. The dimer F^e should be equal to that of two touching spheres, i.e., $(2F^e_{mon} + \Delta F^e_{dim})$. Thus, for 4Dim $\rightleftharpoons$ Oct,

$$\Delta F^e_{(oct-dim)} = F^e_{oct} - 4F^e_{dim} = F^e_{oct} - 8F^e_{mon} - 4\Delta F^e_{dim} \tag{32}$$

Such calculations will become more widely applicable as X-ray studies result in increased knowledge of the geometry of the various states of aggregation of subunit proteins.

In calculating ΔF^e for a molecular expansion process, comparison of Eqs. 13 and 27 shows that the electrostatic free energy of an expanded molecule is much smaller than that of the same molecule in compact form. F^e values calculated by Tanford (46) for hypothetical protein molecules in the compact and expanded forms are compared in Fig. 6 as a function of ionic strength for several values of $\bar{Z}_2$; it is evident that expansion to twice the radius results in a decrease of F^e by close to a factor of 10 at an ionic strength of 0.15; thus, in calculations of the change of electrostatic free energy during an expansion process, it is frequently possible to make the assumption that $\Delta F^e = -F^e_{compact}$. In view of the approximate models used, this assumption is certainly quite reasonable.

IV. BINDING OF HYDROGEN AND OTHER IONS TO PROTEINS—TITRATION CURVES

The charge on a protein molecule is a function of the ionization of its acidic groups, i.e., of equilibria with hydrogen ions, and of the binding by the macromolecule of other ionic species from solution. This subject has been excellently covered recently in two reviews by Tanford (11,55) and the one of Steinhardt and Beychok (17). We will present a few basic concepts and working equations and, since the data on a vast number of systems investigated have been compiled quite comprehensibly and analyzed thoroughly in these reviews, describe their applicability to only a few two examples.

The acid–base titration curve of a protein gives as its primary result number of protons bound to a protein molecule as a function of pH. Information which may be deduced from this is: (1) the type and number of ionizable groups on a protein and thus its charge and charge configuration proton binding as a function of pH, (2) the electrostatic First, protein as a function of pH, and (3) the general gross Excellent state of association. These will be discussed in turn. examples ic theory of titration curves will be summarized. problem are available in the literature and expossible, from some recent reports.

The proton dissociation of any group on a protein is described by Eq. 1 where the product gives a negative charge or is neutral, depending on the nature of the group. The addition of a proton to a molecule which carried an electric charge or removal of one from it requires additional work against the electrostatic free energy, F^e, of the protein; therefore, the apparent dissociation constant of the reaction, K', for each group will be a function also of the total charge. In general, it has been shown that the ionizable groups on a protein can be treated as families of different types, with identical intrinsic dissociation constants, K_i. The free energy, $\Delta F'_n$, of removing a proton from group n on the protein in the presence of all the other charges on the protein is

$$\Delta F'_n = \Delta F_i + \Delta F^e_n \tag{33}$$

where $\Delta F_i = 2.303\,RT\,\log K_i$ is the intrinsic free energy of dissociation of the group, i.e., the free energy change involved in the removal of the proton from group n as such; ΔF^e_n is the change in free energy accompanying the removal to infinity of the n'th proton from the field of all the charged groups on the protein. Since removal of a proton is identical with a change in charge by one group, $\Delta F^e_n = dF^e/d\bar{Z}$. For the reaction $AH \rightleftharpoons A^- + H^+$, setting α as the fraction of groups dissociated and $(1 - \alpha)$ that of groups protonated,

$$K' = \frac{\alpha[H^+]}{(1 - \alpha)} \tag{34}$$

Combining Eqs. 33 and 34

$$\log K_i = \log[H^+] + \log \frac{\alpha}{(1 - \alpha)}$$

or for a single molecule,

$$\text{pH-log} \frac{\alpha}{(1 - \alpha)} = pK_i - \frac{1}{2.303kT}\frac{dF^e}{d\bar{Z}} = pK_i - 0.434\frac{e\psi}{kT} \tag{35}$$

Substituting the value of F^e_n for the spherical smeared charge model within the Debye-Hückel approximation (Eq. 13) gives the Linderstrøm-Lang (18) titration equation per molecule:

$$\text{pH-log} \frac{\alpha}{1 - \alpha} = pK_i - 0.868w\bar{Z}$$

$$w = \frac{\epsilon^2}{2DkT}\left(\frac{1}{b} - \frac{\kappa}{1 + \kappa a}\right) \tag{36}$$

This equation has been quite useful in the interpretation of titration data. A plot of the left-hand side of this equation vs the charge $\bar{Z}$ gives a straight line with a slope of $-0.868w$ and an intercept of pK_i The charge on the protein molecule can be determined from the sum of the number of protons bound or removed from the isoelectric protein and the number of all other ions bound to the protein. These can be determined by a number of techniques which have been discussed elsewhere and will not be discussed here. Equations 35 and 36 are perfectly general and can be used for the analysis of the binding of any ions to the protein, and Eq. 35 can be used to determine experimentally the electrostatic free energy for any protein independently of its shape and charge distribution, as long as no specific expressions for F^e are written out.

The total number of protons bound, $\bar{v}$, summed over all types of groups i with intrinsic dissociation constants $K_i^{(i)}$ is

$$\bar{v} = \sum_{\bar{v}_i} \frac{v[\mathrm{H}^+]^v e^{-e\psi/kT}}{K_i^{(i)}(1 + [\mathrm{H}^+]^v e^{-e\psi/kT}/K_i^{(i)})} \tag{37}$$

where v is the number of groups of type i in the protein. This equation describes exactly the titration curve of the protein if all nonelectrostatic effects on the dissociation of any group are included in K_i. As pointed out previously, the net average charge $\bar{Z}_p$ on a protein depends on $\sum_j v_j z_j$, where z_j is the valence of bound ions of type j.

While Eq. 36 is the form normally used to analyze the titration curves of proteins, its applicability is limited by the same assumption as the calculation of F^e by Eq. 13. Nagasawa and Holtzer (40) have pointed out that the Debye-Hückel approximation always results in straight line relations between pH-log $(\alpha/(1-\alpha))$ and $\bar{Z}$. In practice, such plots usually curve at high charges, $(|Z| > 15)$. Examination of the curvature of such plots for β-lactoglobulin and conalbumin in terms of the empirical parameter $(a - b)$ has resulted in values for the effective mean radius of K^+ and Cl^- ions between 0 and 2.4 Å (40). In the case of β-lactoglobulin, the situation is further complicated by dissociation of the protein into subunits (34,35) in the charge interval examined by these authors; such a dissociation should also lead to curvature in the Linderstrøm-Lang plot.

Tanford (42) has calculated theoretical titration curves for a number of models using various discrete charge distributions and Eq. 19 for F^e. Since these calculations are carried out within the Debye-Hückel approximation, these plots result invariably in straight lines. The intercepts and slopes, however, are strong functions of charge distribution and are dependent on the ionic strength of the surrounding medium. Thus, a variation in reported intrinsic pK's of groups of a single type from protein to protein and with salt concentration for a single protein may be simply the consequences of the assumptions that pK_i and w in Eq. 36 are constant; in this way this equation fails to take all interactions into account.

Other important quantities that may be obtained from a titration curve are the contribution of proton binding to charge fluctuations, the isoelectric point, and the presence of reactions, such as conformational changes or aggregations. Differentiation of Eq. 37 with respect to pH gives

$$\frac{d\bar{\nu}}{d\mathrm{pH}} = -2.303(\overline{Z^2} - \bar{Z}^2) = \overline{\Delta Z_{\mathrm{H}}^2} \tag{38}$$

i.e., the slope of the titration curve at any pH gives a value of the charge fluctuations at that pH. $\overline{\Delta Z_{\mathrm{H}}^2}$ may be calculated from known values of K_i and F^e, since

$$\overline{\Delta Z_{\mathrm{H}}^2} = \sum_{i=1}^{n} [2 + K_i e^{-\epsilon\psi/kT}/a_{\mathrm{H}^+} + a_{\mathrm{H}^+}/K_i e^{-\epsilon\psi/kT}]^{-1} \tag{39}$$

where a_{H^+} is the hydrogen ion activity.

The isoelectric point is simply that pH at which the numbers of positive and negative charges on a protein resulting from the binding of all ions are equal. Variation of this pH with salt concentration has usually been attributed to binding of ions other than hydrogen. Thus, Scatchard and Black (56) have shown that, within the limits of Eq. 13, at the isoelectric point, Eq. 36 leads to the result

$$\mathrm{pH} - \mathrm{pH}_0 = \Delta\mathrm{pH} = -0.868w \sum_i \nu_i z_i \tag{40}$$

where ν_i is the number of salt ions of valence z_i bound to each protein molecule, and pH_0 is the isoelectric pH in the absence of salt. Nagasawa and Holtzer (40), however, have examined this question in terms of the theory of Tanford and Kirkwood (10,42) and have concluded that such shifts in pH may reflect a change in the pK_i of a given group with ionic strength rather than a change in charge due to ion binding. According to the treatment of Tanford and Kirkwood, $pK_i^{(i)}$ at $\bar{Z} = 0$ is not necessarily the true intrinsic dissociation constant of group (i). The true constant $pK_i^{(i)_0}$ is given by

$$pK_i^{(i)_0} = pK_i^{(i)} - 0.434\frac{\epsilon}{kT}\Delta\psi_0 \tag{41}$$

where $\Delta\psi_0$ is the sum of the electrostatic effects of all other charged groups on the group in question at $\bar{Z} = 0$. A change in salt concentration results in changes in ionic screening so that the value of $\Delta\psi_0$ changes also; the direction and magnitude of such a change must be a close function of the charge distribution and exact conformation of the protein since the effects of positive and negative groups in various mutual configurations may conceivably be screened out to different extents.

A. Burial of Groups

Reactions which result either in changes in size, shape, or charge distribution of a protein are reflected frequently in the titration curves. Such reactions may be manifested through changes in the electrostatic interaction parameter, w, in the total number of ionizable groups, n, or in changes in pK_i due to changes in the group environment, such as changes in the distribution of neighboring charged groups and in the dielectric constant of the medium because of burial of the group in the hydrophobic interior of the molecule. This may also result from the participation of the group in a bond, such as an intramolecular hydrogen bond or an intramolecular ion pair or the strong binding of an ion other than hydrogen to the group. The change in pK_i results from a change in the free energy of dissociation of the group in question, and the displacement of the observed pK due to interaction or change in environment is the result of the free energy contribution ΔF_{react} of the additional process:

$$\Delta F_{\text{app}} = \Delta F_i + \Delta F_{\text{react}} \tag{42}$$

This effect has been treated for the cases of the burial of a group inside the protein and of its participation in hydrogen bonds. These will be discussed in turn.

The effect of hydrogen bonding on the titration of an ionizable group has been examined in detail by Laskowski and Scheraga (57), who derived equations for the dependence of the observed ionization constant K'_{obs} on the hydrogen bonding constant, k. If, as before, K' contains the electrostatic interaction term, the observed dissociation constant will be (55)

$$\frac{K'_{\text{obs}}}{[\text{H}^+]} = \frac{\alpha_{\text{obs}}}{1 - \alpha_{\text{obs}}} = \frac{\alpha(1 - \beta)}{1 - \alpha(1 - \beta)} \tag{43}$$

where α is the fraction of nonhydrogen bonded groups dissociated, α_{obs} is the fraction of total groups dissociated, and β is the fraction of groups hydrogen bonded. Now, defining x as $K'/[\text{H}^+]$, the expressions for K'_{obs} for a number of hydrogen-bonded species have been calculated and are listed in Table III. A number of other situations, including cooperative and competing phenomena, are discussed by Laskowski and Scheraga. The principal result of such interactions is a departure of the curve from linearity when plotted according to the Linderstrøm-Lang plot and a

TABLE III
Effect of Hydrogen Bonding on the Dissociation of Ionizable Groups

Species[a]	k	$\dfrac{\alpha_{obs}}{1 - \alpha_{obs}}$	
	$\dfrac{\beta}{(1 - \beta)(1 - \alpha)^2}$	$\dfrac{x(1 + x)}{1 + k + x}$	(44)
$-COOH-----OOC-$ $(-RH-----R-)$	$\dfrac{\beta}{2(1 - \beta)(1 - \alpha)\alpha}$	$\dfrac{x(1 + x)}{1 + 2kx + x}$	(45)
$-COOH----A-$ $-RH----A^- -$	$\dfrac{\beta}{(1 - \beta)(1 - \alpha)}$	$\dfrac{x}{1 + k}$	(46)
$-AH----R-$	$\dfrac{\beta}{(1 - \beta)\alpha}$	$\dfrac{x}{1 + kx}$	(47)
and	$\dfrac{\beta}{(1 - \beta)[(1 - \alpha)^2 + 2\alpha(1 - \alpha)]}$	$\dfrac{x(1 + x)}{1 + k(1 + 2x) + x}$	(48)

[a] R is the ionizable group; A is the hydrogen bond acceptor.

strong deviation of the apparent pK from pK_i, as shown on Fig. 7 for the case of a histidyl-histidyl single bond. On the other hand, knowledge of the displacement of the pK permits the calculation of k if the number and type of hydrogen bonds are known.

The case of the burial of ionizable groups in the hydrophobic interior of a protein molecule has been treated by Tanford (58). In this case, ionization of the group must be preceded or accompanied by a conformational transition. Following Tanford's argument, we take a protein which can undergo a transition from form α to form β, whether the group in question, A, exists in the acid, AH, or basic, A form; The reaction scheme is:

$$
\begin{array}{ccc}
& \xrightarrow{k_1} & \\
\alpha A & \rightleftharpoons & \beta A \\
& \text{I} & \\
K_\alpha \Big\updownarrow & & \Big\updownarrow K_\beta \\
& \text{II} & \\
\alpha AH & \rightleftharpoons & \beta AH \\
& k_0 &
\end{array}
\qquad (49)
$$

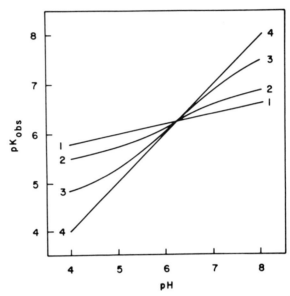

Fig. 7. Observed values of histidyl pK as a function of pH for various strengths of the histidyl–histidyl hydrogen bond (57). Isoelectric point: 5.00; $2w(d\bar{Z}/dpH) = -0.46$; $pK_i^\circ = 6.00$. (1) $k = 0$; (2) $k = 1$; (3) $k = 10$; (4) $k = \infty$.

If $k_0 \neq k_1$, then $K_\alpha \neq K_\beta$, and a group must have an abnormal pK.† The transition from αAH to βA can take place by either path I or II. If form α is more stable in acidic form AH and form β in basic form A, then: $k_0 = [\beta AH]/[\alpha AH] < 1$, $k_1 = [\beta A]/[\alpha A] > 1$. Defining the degree of dissociation of AH as x_A, and the apparent extent of transition as y_β.

$$x_A = \frac{\beta A + \alpha A}{\beta AH + \alpha AH + \beta A + \alpha A}$$

$$y_\beta = \frac{\beta/(\alpha + \beta) - k_0/(1 + k_0)}{k_1/(1 + k_1) - k_0/(1 + k_0)}$$

(50)

Tanford has shown that

$$x_A = y_\beta = \frac{K^*/[H^+]}{1 + K^*/[H^+]}$$

(51)

where $K^* = K_i e^{2w\bar{Z}}$ is the apparent pK of the group in question and is equal to

$$K^* = \frac{k_0(1 + k_1)}{k_1(1 + k_0)} K_\beta = \frac{1 + k_1}{1 + k_0} K_\alpha$$

(52)

† For the case in which $k_0 = k_1$, $K_a = K_\beta$ and the ionization of the group is not related to the transition, i.e., its environment does not change and the case is trivial.

For the case where conversion is essentially complete at the two ends of the ionization of group A, i.e., $k_0 \sim 0$, $k_1 \sim \infty$, and if the buried group is acidic, K_β is its normal dissociation constant, Eq. 52 reduces to

$$pK^* = pK_i + \log\left[1 + \frac{1}{k_0}\right] \qquad (53a)$$

while, if the uncharged buried group is basic, the equation is

$$pK^* = pK_i - \log(1 + k_1) \qquad (53b)$$

In Eqs. 53a and b, $1/k_0$ and k_1 are equilibrium constants in the direction of the state in which the group would be normally uncharged. Such burial can result in a large displacement of the pK.

When the transition is associated with the ionization of more than one group, the relations between K^*, x, and y become considerably more complex; x_A can differ from y_β, and if the normal pK's of the groups involved are quite different, the transition curve can take on the appearance of a two-stage reaction. pK displacements of several pH units are known to occur in a number of proteins. One may cite the buried imidazole groups of hemoglobin (59) and human carbonic anhydrases B and C (60,61), the buried tyrosines of a number of proteins (62–64) and the buried carboxyls of the β-lactoglobulins (65–69), as well as the anomalous imidazole of β-lactoglobulin C (68,70). As typical examples, let us examine the cases of carbonic anhydrase and β-lactoglobulin C.

The titration curve of carbonic anhydrase C at $\Gamma/2 = 0.15$ and 25°C is shown in Fig. 8. The meaning of the symbols and the curves are explained in the legend. Considering only the region of acid pH, the forward and reverse titration curves are widely separated between pH 4 and 7. The difference in protons bound between the two curves is shown in the inset and is seen to reach 7 groups at pH 5 and then to drop rapidly to zero. These groups, which become protonated with an apparent pK of 4.5 in the forward titration, dissociated in the back titration with a pK of 6.3 which is normal for imidazoles. A conformational examination of the same protein (71,72) has shown that a molecular unfolding occurs in the same pH region. Thus, the change in the apparent pK of the groups can be related to the unmasking of previously buried groups. Just as in hemoglobin, the acid unmasking of the histidine residues is irreversible, so that it does not lend itself to the treatment in terms of Eqs. 49–53. It is interesting, however, to point out that for a hypothetical reversible reaction with such values of histidine pK's as are found in carbonic anhydrase C, application of Eq. 53 would result in $k_1 \sim 65$, i.e., a free energy of group burial of ~ -2.5 kcal/mole. While this value is not applicable to carbonic anhydrase, it may serve as an illustra-

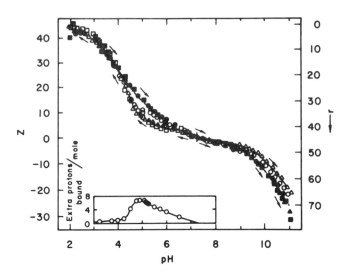

Fig. 8. Titration curves of human carbonic anhydrase C at $\Gamma/2 = 0.15$ and $25°$. The assumed molecular weight of the protein is 30,000. The open symbols indicate the initial titration starting at the isoionic point, pH 7.3. The solid symbols indicate back-titration with KOH, starting at pH 1.85 to 1.9. The circles indicate a lyophilized preparation, whereas the triangles and squares are for different crystalline preparations. The solid and dashed lines denote theoretical curves. The dashed line is for the back-titration after exposure to pH 2 or below. The inset is the difference curve illustrating the extra hydrogen ions bound on reverse titration. Taken from (*61*).

tion of the contribution which group burial may make to the instability of a globular protein molecule.

The titration curve of β-lactoglobulin C (*68*) represents a particularly interesting case since it is characterized by two reversible transitions in the pH regions of 7.5 and 5.0. In the first, a carboxyl group, previously unavailable to ionization, becomes exposed to solvent; in the second, a basic group which is unprotonated in the low pH region becomes available to protonation. The experimental results are shown in Fig. 9. The titration curve is mostly quite normal and can be analyzed in terms of a *w* value of 0.04 and the number of ionizable groups with corresponding p*K*'s listed in Table IV. The two anomalies are evidenced by a steepening of the curve between pH 7 and 8.5 and a flattening in the region between pH 4.5 and 6.0. Theoretical curves calculated on the basis of $w = 0.04$ and varying numbers of ionizable groups with normal p*K*'s show that the pH 7.5 transition is accompanied by a shift from 48 carboxyls (per two chain dimer) to 50 carboxyls; the apparent p*K* of these groups, p*K**, is 7.25, i.e., this transition is identical with the one reported

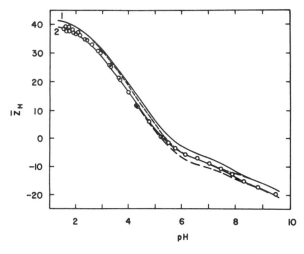

Fig. 9. Calculated titration curves of β-lactoglobulin C (*68*). Circles are experimental points. Upper solid line: calculated for 46 side chain carboxyls and 6 imidazoles titrating normally. Short-dashed line: 48 side chain carboxyls and 6 imidazoles titrating normally. Long-dashed line: calculated for 48 side chain carboxyls and 4 imidazoles titrating normally with 2 imidazoles buried and unavailable to titration. Lower solid line: 48 side chain carboxyls and 4 imidazoles titrating normally, 2 imidazoles buried below pH 4.5. Both solid lines include the 2 abnormal carboxyls with apparent $pK = 7.25$.

for variants A and B (*65,66*). In the pH 5.0 region, the transition is from a curve calculated on the basis of 40 basic residues becoming protonated to one with 42 such residues. Both transitions are found to be reversible and to be accompanied by reversible changes in the optical rotation of the protein, indicating a small conformational change (since *w* remains essentially constant). These two transitions have been analyzed in terms of Eqs. 49–53. For the pH 7.5 transition, application of Eq. 53a results

TABLE IV

Groups Titratable below pH 9 in β-Lactoglobulin C (35,500 MW)

	Number of groups		pK
α-COOH	2	(2)	3.75
β,γ-COOH (normal)	48		4.66
β,γ-COOH (anomalous)	2	(50)[a]	7.25
Imidazole (normal)	4		7.25
Imidazole (anomalous)	2	(6)	
α-NH$_2$	2	(2)	7.80
Total cationic		(42)	

[a] The number in parentheses is the number of residues found in amino acid analysis

in $k_0 = 2.5 \times 10^{-3}$ and $\Delta F^0 = 3.5$ kcal/mole (73,70). Examination of the transition as a function of temperature, resulted in a value of 8.0 kcal/mole for ΔH^*. However, since ΔH^* encompasses the enthalpies of all the processes involved in the transition

$$\Delta H^0 = \Delta H^* - \sum_i \Delta H_i^0 \tag{54}$$

where ΔH^0 is the enthalpy of the transition and ΔH_i^0 are the standard heats of dissociation of the ionizable groups involved. Since a single carboxyl with $\Delta H_i^0 = 1.0$ kcal/mole is involved, $\Delta H^0 = 7$ kcal/mole and $\Delta S^0 = 12$ e.u.; from these values Tanford and Taggart (73) have concluded that the buried carboxyl is most likely hydrogen bonded in the interior of the molecule.

A similar analysis of the pH 5.0 transition is difficult since the key group involved in the transition must be a carboxyl, and $pK^* = 5.25$ is close to $pK_i = 4.65$ (68). Equation 53 results in k_0 between 0.143 and 0.333 since k_1 must have values between 1 and ∞. The corresponding free energy of the transition, ΔF^0, is ~ 1.0 kcal/mole. The most noteworthy feature of this transition, however, is the removal from contact with solvent of one basic group as the pH is decreased. This may be the result of one of two mechanisms: (1) A cationic residue is drawn into the hydrophobic interior during the conformational change and is forced to dissociate and become uncharged; (2) the transition is accompanied by the formation of an ion pair between a cationic group and a carboxylate ion, with the removal of this ion pair from contact with solvent. Jacobsen and Linderstrøm-Lang have shown that such an ion pair would not be stable in contact with water (74). The thermodynamics of the burial process were examined as follows (68): ΔH^* was found to be zero in β-lactoglobulin C; in variant B, in which the transition does not involve the burial of a basic group, ΔH^0 was measured to be -5.1 kcal/mole. Then, according to Eq. 54, $\Delta H_{basic} = \Delta H^* - \Delta H^0 - \Delta H_{COOH}^0 = 4.1$ kcal/mole. The corresponding entropy and free energy are: $\Delta S_{basic} = -14$ e.u. and $\Delta F_{basic} \sim 0$ kcal/mole. This corresponds to a structure destabilizing free energy contribution of 9 and 6.5 kcal/mole for the two mechanisms since in the buried state the pK of the key group has an effective value of zero. These values are consistent with the burying process being predominantly of a hydrophobic nature.

An interesting way of characterizing the transition of a protein from titration data has been described by Perlmann et al. (75,76). In this approach, the buffering capacity of the protein, $\beta' = (\partial h/\partial pH)_{T,p}$, where h, the number of neutralized groups per protein molecule, is plotted as a function of pH. A maximum is obtained in this curve at the point of a transition.   Integration over the buffering range gives Δh, i.e., the number of groups exposed during the conformational change, since

$$\int_{\mathrm{pH_1}}^{\mathrm{pH_2}} \left(\frac{\partial h}{\partial \mathrm{pH}}\right)_T d\mathrm{pH} = \Delta h \tag{55}$$

The change of pH with temperature of the maximum position of β' results in the entropy and enthalpy of the transition, since

$$\Delta \bar{S} = 2.303 RT \Delta \bar{n}_{\mathrm{H^+}} \left(\frac{\partial \mathrm{pH}}{\partial T}\right)_{\mathrm{tr}}$$

$$\Delta \bar{H} = 2.303 RT^2 \Delta \bar{n}_{\mathrm{H^+}} \left(\frac{\partial \mathrm{pH}}{\partial T}\right)_{\mathrm{tr}} \tag{56}$$

where $\Delta \bar{S}$ and $\Delta \bar{H}$ are the "melting" entropy and enthalpy per protein molecule and $\Delta \bar{n}_{\mathrm{H^+}}$ is the number of protons per protein molecule involved in the transition.   The heat of melting per protein molecule is $\Delta \bar{\bar{H}} = \Delta \bar{H}/\Delta \bar{n}_{\mathrm{H^+}}$.   The data of Perlmann et al. on the conformational change in pepsinogen are shown in Fig. 10.   Analysis of these data results in the involvement of three histidine residues in the transition independently of temperature between 30 and 60°C and of ionic strength between 0.01 and 0.2. $\Delta \bar{\bar{H}}$ was found to be 50 kcal/mole.

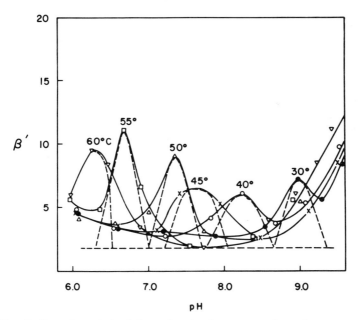

Fig. 10.   Buffer values derived from the titration curves of pepsinogen at various temperatures.   The dashed lines under each curve indicate the pH limits used for the intergration (75).

B. Intermolecular Associations

In the case of protein associations such as $nM \rightleftharpoons A$, the number of protons bound per unit mass changes. Such a change can be the result of two factors: (1) The electrostatic free energy is a function of the size and charge distribution of the molecule, so the parameter $w = \bar{F}^e/\bar{Z}^2 kT$ will change as

$$w_A = w_M \frac{[1 + n^{1/3}\kappa(a - b)](1 + \kappa a)}{[1 + \kappa(a - b)](n^{1/3} + n^{2/3}\kappa a)} \tag{57}$$

(2) Specific groups on the protein molecule may be involved in the intermolecular bonds formed with the result that their pK's become displaced. From such changes in w and the pK's of groups, i.e., from changes in the number of protons bound, it is possible to gain information on the thermodynamic parameters of the association, just as in the case of intramolecular hydrogen bonding or group burial.

Lebowitz and Laskowski (77) have shown that when a protein association is accompanied by the release or uptake of protons, the equilibrium constant of the reaction can be determined from the titration curve. For the reaction†

$$A + B \rightleftharpoons AB + \bar{q}[H^+]$$

$$K_{app} = \frac{[AB]}{[A][B]}$$

$$\bar{q} = \frac{d \log K_{app}}{dpH} \tag{58}$$

From a plot of the moles of H^+ evolved as a function of concentration of protein B added to a constant amount of protein A, both $\bar{q}$ and K_{app} can be determined.

Integration of the last of Eq. 58 gives

$$(\log K_{app})_2 - (\log K_{app})_1 = \int_{pH_1}^{pH_2} \bar{q} \, dpH \tag{59}$$

Thus, if the equilibrium constant at one pH is known, it is possible to obtain the value at any other pH simply by determining the number of protons released and integrating under the curve of $\bar{q}$ vs pH. By applying this technique to the interaction of trypsin with soybean trypsin inhibitor, Lebowitz and Laskowski (77) have obtained the results shown in Table V, where the values of K_{app} are compared with similar data obtained from the completely independent technique of gel filtration (79).

---

† Laskowski (77a) has also shown that an association reaction $nA \rightleftharpoons A_n + mH^+$ can be treated in a similar manner.

TABLE V

Equilibrium Constants for the Formation of the Trypsin-Soybean
Trypsin Inhibitor Complex, K (mole-liter^{-1} $\times$ 10^7)

pH	K (potentiometric technique) (77)	K (gel filtration)^a (78,78a)
4.25	10.5	10.8
4.50	2.3	2.8
4.75	0.7	0.5
5.50	0.0	0.0

^{a} Data of Nichol and Winzor (78) corrected by Gilbert (78a).

These two sets are essentially identical, indicating that titration techniques
may be used to great advantage in the characterization of interacting
systems.

In a series of papers on the conversion of fibrinogen to fibrin, Scheraga
and coworkers (79–81) have shown how to use the changes in pH which
accompany the reaction to calculate the thermodynamic parameters of
bonds between ionizable groups which result in polymerization (82,83).
For the reaction between pairs of groups, one in the acid form, G_1H, the
other in the basic form, G_2,

$$G_2 + G_1H \rightleftharpoons G_2G_1H \qquad K_{assoc}$$
$$G_1H \rightleftharpoons G_1 + H^+ \qquad K_1$$
$$G_2H \rightleftharpoons G_2 + H^+ \qquad K_2 \qquad (60)$$

The fraction, x_{ij}, of the r such pairs per protein monomer which become
bonded to each other in the $nM \rightleftharpoons P$ reaction is

$$x_{ij} = \frac{K_{assoc}}{1 + K_{assoc} + K_1/[H^+] + [H^+]/K_2 + K_1/K_2} \qquad (61)$$

while the net number of protons, q, released per pair ij during polymerization is

$$q = x_{ij} \frac{[H^+]/K_2 - K_1/[H^+]}{1 + K_1[H^+] + [H^+]/K_2 + K_1/K_2} \qquad (62)$$

where $K_i = K_{i,int} \, e^{2w\bar{Z}}$.

The quantity measured experimentally, Δh, i.e., the difference between
the polymer and the monomer of the number of protons bound per mono-
mer protein, is evidently $\Delta h = rq$. It is evident that Δh and q will be
zero at a pH which corresponds to $[H^+]_0 = (K_1K_2)^{1/2}$. Furthermore, Δh
attains maximum and minimum values at pH's which correspond to

$$[H^+]_{max} = K_2[(1 + K_{assoc} + K_1/K_2)(1 + K_1/K_2)]^{1/2}$$
$$[H^+]_{min} = K_1/[(1 + K_{assoc} + K_1/K_2)(1 + K_1/K_2)]^{1/2} \quad (63)$$

and the magnitudes of Δh at the maximum and minimum are

$$\Delta h_{max} = -\Delta h_{min} = \frac{rK_{assoc}}{[(1 + K_{assoc} + K_1/K_2)^{1/2} + (1 + K_1/K_2)^{1/2}]^2} \quad (64)$$

The heat, ΔH_{assoc}, evolved per mole of monomer during polymerization is

$$\Delta H_{assoc} = r(x_{ij}\Delta H^0_{assoc} - q_1\Delta H^0_1 + q_2\Delta H^0_2) \quad (65)$$

where ΔH^0 is the enthalpy of the bond formation, q_1 is the number of protons absorbed by G_1 per ij'th pair, q_2 is the number of protons released by G_2H per ij'th pair, and ΔH^0_1 and ΔH^0_2 are the standard enthalpies of of ionization of the respective groups.

Scheraga and coworkers (79–81) have applied this theory to the conversion of fibrinogen to fibrin. Their experimental results, expressed as protons evolved per 10^5 g of fibrin, are shown in Fig. 11. A maximum occurs at pH 6.4 and the data cross zero at pH 7.6. Analysis of the data in terms of Eqs. 60–65 resulted in values of $\Delta h_{max} \sim 1.0$ and $r = 4$–6.

While in the previous example the reaction in question had gone to completion, a frequent situation is one in which the system is in equilibrium, $nM \rightleftharpoons P$. If the titration curve of the monomer is known, that of the polymer may be derived from equilibrium data on the reacting system. Furthermore, if the geometric structures of the monomer and polymer are

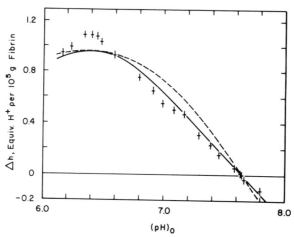

Fig. 11. Uptake of hydrogen ions, Δh, during conversion of fibrinogen to fibrin plotted as a function of the initial pH. Theoretical curves are calculated according to Eq. 62. Solid line: calculated for $r = 6$. Dashed line: calculated for $r = 4$. $\Delta h_{max} = 0.97$ (81).

known, the electrostatic free energy change contribution during the polymerization may be calculated. In such a system, at any pH, the difference between the number of protons bound per monomer, Δh_{exp}, is measured directly from the difference between the titration curves. If the polymerization is induced by a temperature change, the necessary quantities may be obtained simply by changing the temperature. Then, the number of protons released due to the reaction, Δh_{react}, is obtained by correcting Δh_{exp} for the number of protons released, $\Delta h^{(\Delta H_i)}$, due to the enthalpy of ionization, ΔH_i of the groups i titrating at the given pH, i.e.,

$$\Delta h_{react} = \Delta h_{exp} - \sum_i \Delta h^{(\Delta H_i)} \tag{66}$$

The quantity Δh necessary for use with equations such as 60–65 is simply

$$\Delta h = \Delta h_{react}/x_{pol} \tag{67}$$

where x_{pol} is the weight fraction of the protein polymerized, and the charge on a polymerized molecule, $\bar{Z}_{pol}$, is

$$\bar{Z}_{pol} = n(\bar{Z}_{mon} + \Delta h/x_{pol}) \tag{68}$$

This analysis has been applied to the tetramerization of β-lactoglobulin A from the isoelectric dimer to an octamer, $4A_2 \rightleftharpoons (A_2)_4$, (33,36,39,84). This system is known to be in rapid equilibrium, it has a maximal free energy of association at pH 4.4–4.7, and it has a negative enthalpy. Therefore, titration curves were obtained (37) at 2.4° where tetramerization is extensive and 28° where it is essentially negligible. Their difference gives Δh_{exp}. $\Sigma \Delta h^{(\Delta H_i)}$ was obtained directly from the titration curves at the same temperature of β-lactoglobulin B which polymerizes to a negligible extent. x_{pol} was calculated from known equilibrium constants, so that $\bar{Z}_{pol}$ could be obtained directly. The pertinent results are shown in Fig. 12a. Since the polymer is known to be a compact structure of 422 symmetry (39), it could be approximated by a sphere and its electrostatic free energy was calculated within the Linderstrøm-Lang approximation, using Eq. 13, and ΔF^e of the polymerization reaction was calculated as described above (see Section III.B.4). The results, shown in Fig. 12b, indicate that as pH decreases and the protein assumes a progressively increasing positive charge, a strong electrostatic repulsion sets in and contributes to the dissociation of the octamer, since $\Delta F^0_{exp} = \Delta F_{app} + \Delta F^e$, where ΔF_{app} is the nonelectrostatic observed apparent free energy of the polymerization reaction and ΔF^0_{exp} is the apparent standard free energy experimentally observed. This analysis has shown that the octamer formation of β-lactoglobulin A is accompanied by the forced protonation of four carboxylic residues per 18,000 dalton single chain subunits, i.e., four

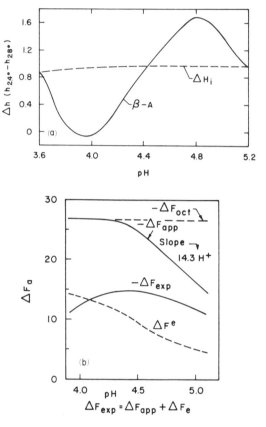

Fig. 12. (a) Uptake of hydrogen ions during formation of the β-lactoglobulin A octamer from the dimer; the dashed line represents the proton uptake due to temperature lowering (ΔH_i). (b) Contributions of the electrostatic free energy change, the conformational change, and the intrinsic free energy of octamer formation to the observed free energy in the β-lactoglobulin A system (*85*).

groups per site of association or a total of 32 groups. The availability of such information is required if the system is to be analyzed further in terms of the mechanism of the association.

The β-lactoglobulin A system may serve furthermore as an example of the coupling of an association described by Eqs. 60–68 and 29–32 and a conformational transition described by Eqs. 49–54 (*85*). While the decrease in observed free energy of tetramerization below pH 4.4 can be explained strictly in terms of the change in electrostatic free energy as described above, it would appear that the similar decrease above pH 4.5 may be related to a conformational transition. This protein is known to undergo a small conformational change centered around pH 5.0 at 25°C

(70). Analysis in terms of Eqs. 49–54 has shown that per dimer this change follows the reaction $\alpha_2 + 4H^+ \rightleftharpoons (\beta H_2)_2$ $(K_1 = [(\beta H_2)_2]/[\alpha_2][H^+]^4)$. If only the acid form $(\beta H_2)_2$ can tetramerize to the octamer, $K_{oct} = [((\beta H_2)_2)_4]/[(\beta H_2)_2]^4$. The values of ΔF^0_{app}, however, were obtained from molecular weight measurements which do not distinguish between forms α_2 and $(\beta H_2)_2$ of the protein. Then,

$$K_{oct} = \frac{K_{app}(1 + K_1[H^+]^4)^4}{K_1^4[H^+]^{16}} \qquad (69)$$

Setting $K_1 = 10^{-9}$ and using the values of K_{app} calculated above, K_{oct} was found to be constant over the pH range studied and resulted in $\Delta F^0_{oct} = -27$ kcal/mole. The contributions to ΔF_{exp} of the various factors determined from titration data are shown in Fig. 12b.

V. INTERMOLECULAR INTERACTIONS

The discussion in the previous two sections was limited essentially to the electrostatic free energy of a protein, methods of determining it and its dependence on various factors, such as properties of the medium and size and shape of the protein molecule, and interactions between groups on a single molecule and between different protein molecules in solution. In this section the electrostatic contributions to protein interactions will be considered. In general, interactions between proteins involve the cooperative effects of various types of attractive and repulsive forces (86), and it is quite difficult even to attempt to unravel their contributions. Nevertheless, the contributions of some electrostatic forces, in particular, may be estimated, and some methods involved in this will be described. These will include short range (intermolecular complex formation) and long range (virial effects) interactions.

A. Complex Formation

When complexes are formed between two protein molecules, the total free energy of the association, ΔF_{assoc}, is given by:

$$\Delta F_{assoc} = \Delta \bar{F}^e_{mol} + \Delta \bar{F}^e_{site} + \Delta \bar{F}^{n-e} \qquad (70)$$

where $\Delta \bar{F}^e_{mol}$ is the electrostatic free energy of interaction between the entire molecules, $\Delta \bar{F}^e_{site}$ is the electrostatic free energy between the touching sites, and $\Delta \bar{F}^{n-e}$ is the free energy contribution of all nonelectrostatic factors, such as hydrophobic interactions, hydrogen bonding, and conformational (entropic) and dispersion forces. Depending on the charge and charge distribution of the molecules, ΔF^e_{mol} may be attractive (if the interaction is between molecules of opposite charge) or repulsive [if the interaction is between molecules of like charge, such as would be the case of the dimerization of a protein, e.g., lysozyme (87)]. $\Delta \bar{F}^{e}_{\text{site}}$ will be attractive (if the association involves the interaction between groups of opposite sign) and repulsive (if charges of identical sign are present on the two interacting sites). We shall limit our discussion to polymerization and, more specifically, to dimerization where the bond is electrostatic between charged groups of opposite sign, such as is shown in Fig. 13, where two molecules of identical mean net charge, $\bar{Z}$, associate at a site with the formation of two bonds between a positive and a negative charge and repulsion between a pair of groups of identical sign. The various electrostatic contributions within this simple model, superimposed on the Linderstrøm-Lang approximation, can be expressed as a first approximation as the sum of screened Coulombic potentials with a general Verwey-Overbeek repulsion between the two touching spherical molecules. ΔF^{e} for the $2A \rightleftharpoons A_2$ reaction is:

$$\Delta F^{e} = \psi_0^2 \frac{Db}{2} + \frac{\epsilon^2}{DR_1} e^{-\kappa R_1} \sum_{ij} Z_i Z_j \qquad (71)$$

where ψ_0 is the surface potential on the molecule given by Eq. 13, D is the dielectric constant of the medium in the domain of interest, b is the radius of the molecule, R_1 is the distance between the charges on sites 1 and 2, ϵ is the electronic charge, κ is the Debye-Hückel parameter, and Z_i and Z_j are the charges on each interacting point on the site. Since Z_i and Z_j will be functions of the respective dissociation constants when a protein–

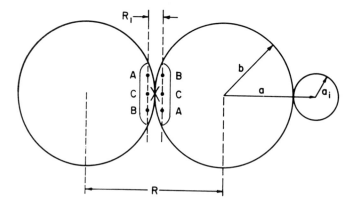

Fig. 13. Model for the electrostatic contribution to protein dimerization: Two A-B attractions, one C-C repulsion between sites separated by distance R_1. X indicates the position of the dyad axis of symmetry perpendicular to the plane of the paper.

protein association is examined as a function of pH, we can write for each attraction between a cationic and anionic group,

$$\Delta F_a = - \left(\frac{K_a e^{2wZ}}{[\text{H}^+] + K_a e^{2wZ}} \right) \left(\frac{[\text{H}^+]}{[\text{H}^+] + K_c e^{2wZ}} \right) \frac{\epsilon^2 e^{-\kappa R_1}}{D R_1} \tag{72}$$

where K_a and K_c are the dissociation constants of the anionic and cationic groups in question; for a repulsion, the free energy contribution is given by the product of two terms involving K_a or K_c, depending on whether the groups are both anionic or both cationic.

A calculation of this type has been carried out for the dimerization of α-chymotrypsin (87a). From the pH dependence of sedimentation data, Egan et al. (88) had postulated that the association occurs between two identical sites, each of which contains a cationic and an anionic residue which interact in pairs, as well as anionic groups which repel each other in pairs. Such an interaction is perfectly plausible if the pair of mutually repelling anionic groups are located along a diameter; in this way, a dyad axis of symmetry can be maintained. This situation is depicted on Fig. 13. Calculations were carried out using Eq. 72 and the known titration curve of this protein (89). Using pK values of 3.3 and 8.5 for the attractive pairs and 5.0 for the repulsive interaction, ΔF^e for the reaction was obtained as a function of pH. It is compared with the experimental pH dependence of the free energy† in Fig. 14a. The results indicate that such a mechanism is quite plausible. Furthermore, the pK's of the three groups are correct for α-carboxyl (or an aspartate carboxyl), α-amino, and a side chain carboxyl. It should be emphasized, however, that the agreement between such a calculation based on a model and experimental data does not prove the validity of a mechanism for the reaction in question; it only establishes that the mechanism is possible and must not be eliminated on the basis of the available evidence. In fact, present knowledge of the space coordinates of the α-chymotrypsin dimer in crystalline state (91a) renders possible the assignment of the interacting pairs of ions to histidine 57 and the α-carboxyl of tyrosine 146; there is no pair of mutually repelling carboxyl residues in the site of interaction. The thermodynamic data obtained between pH 3 and 5 are absolutely consistent with this symmetrical pair of interactions (91b). The association behavior of α-chymotrypsin at higher pH values must reflect other interactions on the molecular level. As another example, in the case of β-lactoglobulin A tetramerization, the bell-shaped pH dependence could be accounted for with equal success by a combination of electrostatic repulsion with a conformational

† This was calculated from Egan's sedimentation data using the Gilbert theory (90,91).

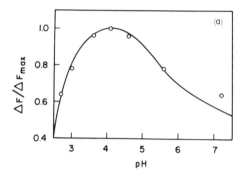

Fig. 14a. Comparison as a function of pH of the experimentally measured free energies of dimerization of α-chymotrypsin (circles) with calculated values, as described in the text. The free energies are expressed as $\Delta F/\Delta F_{max}$ since the absolute values may be changed by changing the value of the dielectric constant between the two spherical molecules (*87a*).

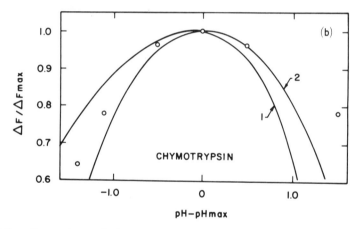

Fig. 14b. Comparison of the experimental data on chymotrypsin dimerization (circles) with curves calculated using the fluctuating charge model. The meaning of curves 1 and 2 is explained in the text (*92*).

change, described above, the fluctuating charge mechanism (*92*), described below, and a model based on the ionization of two specific groups with different pK's. Thus, in order to establish an interaction mechanism, additional information is required.

Other mechanisms by which the proteins may complex as a result of their electrolytic nature involve charge–dipole and dipole–dipole interactions, the free energy contributions of which are given by the relations if the distance between the two groups is large compared to the distance between the two ends of the dipole:

Charge–dipole:
$$\Delta F = \frac{Z\epsilon\mu\ \cos\gamma}{DR^2} \tag{73a}$$

Dipole–dipole:
$$\Delta F = \frac{\mu^2\ \cos\gamma}{DR^3} \tag{73b}$$

where μ is the dipole moment of the interacting groups, R is the distance between the two groups, and γ is the angle formed between the dipole on group 1 and the vector in the direction of the charge in group 2 in case (a) and the angle between the two dipole in case (b).

A rather interesting type of electrostatic interaction is that due to the fluctuation of charges. Kirkwood and Shumaker state that ". . . steric matching of a constellation of basic groups on one molecule with a complementary constellation on the other could conceivably produce a redistribution of protons leading to a strong and specific attraction. . . ." Hill (*93*) has examined this possibility and his treatment follows.

The interaction, inter- or intra-molecular, is considered as taking place between independent sites as depicted in Fig. 15. Therefore, a single pair of such interacting sites are considered as a system in a grand ensemble. Let j_1 and j_2 be the partition functions of molecules or ions bound at sites 1 and 2 of a pair; let $V_{AA}\ (R)$, $V_{AB}\ (R)$, $V_{BA}\ (R)$, and $V_{BB}\ (R)$ be the free energies of interaction between the sites (or groups) separated by a distance R when both sites are occupied (AA), when site 1 is occupied and site 2 unoccupied (AB), the reverse (BA), and when both sites are unoccupied (BB). The potential zero is chosen at $r = \infty$ in each case. The potential of average force $W(R)$ between the two sites is:

$$e^{-W/kT} = \frac{e^{-V_{BB}/kT} + j_1\lambda e^{-V_{AB}/kT} + j_2\lambda e^{-V_{BA}/kT} + j_1 j_2\lambda^2 e^{-V_{AA}/kT}}{1 + (j_1 + j_2)\lambda + j_1 j_2\lambda^2} \tag{74}$$

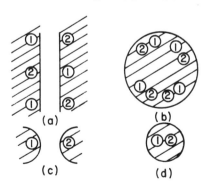

(a)   (b)

(c)   (d)

Fig. 15.   (a) Matching pairs of interacting sites on two large molecules.   (b) Pairs of interacting sites in the same large molecule. (c) Pair of interacting sites on two small molecules.   (d) Pair of interacting sites in the same small molecule (*93*).

where $\lambda = e^{\mu/kT}$ and $\mu = \mu^0 + kT \ln c$; μ is the chemical potential of the bound molecules or ions, and c is their concentration. Defining the probability θ_i that a given site of type i is occupied and introducing the dissociation constants, K_i, $j_i\lambda = c/K_i$. In the case of proton binding, this reduces to $[H^+]/K_{int} = 1/x$. It should be noted that in Eq. 74 the standard state is that with the groups dissociated. Furthermore, $K_1 j_1 = K_2 j_2$. Hill defines a correlation coefficient ξ in such a manner that at $\xi = 1$ the maximum attractive force exists between the matching constellations, and he applies this criterion to two cases.

Case 1: The unoccupied sites (B) are equivalent ($j_1 = j_2$, $K_1 = K_2$) and have a charge of -1. The ion being bound has a charge of -2. The pairs of interacting charges are BB: $--$, AB: $+-$; BA: $-+$, AA: $++$, leading to two attractive and two repulsive potentials.

Case 2: Unoccupied site 1 has zero charge, unoccupied site 2 has a charge of -1 ($j_1 \neq j_2$, $K_1 \neq K_2$). The ion bound has a charge of $+1$. The pairs are BB: $0-$, AB: $+-$; BA: 00, AA: $+0$, leading to one attractive potential.

Examination of these two cases in terms of ξ over a number of conditions leads Hill to conclude that "the Kirkwood–Shumaker proposal that matching constellations might lead to an essentially frozen optimal bound ion distribution is a reasonable one," as the bound ions tend to distribute themselves for optimal attraction between sites. Calculations in terms of Eq. 74 result in values of W/kT as high as -2 for reasonable potentials V_{AB}, indicating that the fluctuation of charges may lead to a significant attraction between specific sites on the molecules.

This treatment has recently been extended to the case of two identical basic groups (such as carboxylates) which can bind an identical ion (such as a hydrogen ion) (*92*). In this case the standard state for Eq. 74 was taken as that in which both groups are associated (the proper form for W is obtained by dividing Eq. 74 by $j^2\lambda^2$). The possible group pairs are BB: $--$; AB: $0-$; BA: -0; AA: 00. The corresponding interactions are V_{BB} a charge–charge repulsion, $V_{AB} = V_{BA}$, a charge–dipole attraction given by Eq. 73a and $V_{BB} = 0$. Here again, calculation of ξ indicates that the proton distribution will tend to be frozen in an optimal manner for attraction. Calculation of W again shows that this mechanism may result in an attractive force of the order 1–2 kcal/mole. The pH dependence of this attraction displays a maximum near the pK of the group in question; it falls off on both sides, having significant values over a range of one pH unit. A typical curve is shown in Fig. 14b where pK_2 = pK_1 + 2 (curve 1). In curve 2 the pH dependence is calculated by taking into account electrostatic interaction, i.e., by setting $x = K_i e^{2w\bar{Z}}/[H^+]$,† where $w\bar{Z}$ is

---

† This equation was written incorrectly in the original paper.

from the chymotrypsin titration curve. As can be seen, this mechanism can reproduce a pH dependence of protein dimerization as actually found in experiment. It is interesting to recall that an equally good fit of the data was obtained with a totally different mechanism (see p. 43).

B. General Interactions

In addition to interactions that lead to specific complex formation, the electrostatic interactions between globular proteins may result simply in an activity coefficient effect which may reflect both attractive and repulsive forces. Several such effects will be treated. In general, an intermolecular interaction can be expressed simply by a change in the activity coefficients or chemical potentials of the interacting species, i.e., $(\partial \mu_i^{(e)}/\partial C_j)_{T,p} = RT \,(\partial \ln \gamma_i/\partial C_j)_{T,p}$, where C is the concentration, γ is the activity coefficient, and $\mu^{(e)}$ is the excess chemical potential, defined as $\mu = RT \log C + \mu^{(e)} + \mu^0 \,(T,p)$. Conclusions on the formation of complexes are reached by interpreting the significance of the changes in the activity coefficient.

1. Virial Coefficients

Using the cluster integrals of the McMillan–Mayer theory (94), Stigter and Hill (95) have calculated the second and third virial coefficients of charged colloidal spheres surrounded by a Gouy–Chapman ionic double layer. This model is similar to the Linderstrøm-Lang model of proteins. Writing out the osmotic pressure, π, in the form of a virial expansion in ρ, where ρ is the number of colloidal particles per unit volume, Stigter and Hill (95) obtain:

$$\frac{\pi}{kT} = \rho + B_2\rho^2 + B_3\rho^3 + \cdots$$

$$B_2 = -\frac{1}{2V} \int_v \int_v (e^{-W_{12}/kT} - 1)\, dr_1\, dr_2$$

$$B_3 = -\frac{1}{3V} \int_v \int_v \int_v (e^{-W_{12}/kT} - 1)(e^{-W_{23}/kT} - 1)(e^{-W_{13}/kT} - 1)\, dr_1\, dr_2\, dr_3$$

$$(75)$$

where W_{ij} is the potential of average force between two spherically symmetrical colloidal particles in positions r_i and r_j, respectively. The potential of average force $W(R) = F^e$ between two spherical particles at large distances was taken as the Verwey–Overbeek potential (Eq. 31). The results of their calculations of B_2 are given in Table VI, where they are presented in corresponding states with $\kappa^3 B_2$ as a function F, where

TABLE VI
Calculations of the Second and Third Virial
Coefficients of Charged Spheres (95)

F	$\kappa^3 B_2$	$\kappa^6 B_3$
0.5	2.86	
1	5.38	
2	9.8	
5	21.2	
10	36.1	338
20	57.5	1,014
50	102	3,830
100	150	9,330
200	216	
300		31,700
500	332	
1,000	447	100,700
2,000	589	
3,000		248,000
5,000	825	
10,000	1,040	596,000

$$\kappa^3 B_2 = 2\pi \int_0^\infty (1 - e^{-W(x)/kT}) x^2 \, dx$$

$$\frac{W(x)}{kT} = F \frac{e^{-x}}{x}$$

$$F = \frac{D\kappa b^2}{kT} \left(\frac{\psi_0}{f} \right)^2 e^{2\kappa b} \tag{76}$$

In these equations all the symbols have their previous meaning and f is a complicated parameter, calculated by Hoskins (96). Stigter (97) further extended this treatment by discussing F^e in terms of the Donnan, Debye–Hückel, and Gouy–Chapman models.

2. Isoionic Proteins; Charge Fluctuations

As has been pointed out above, in globular proteins the number of basic sites generally exceeds the number of protons bound per molecule. This results in the possibility of many configurations of the protons on the protein, differing little in free energy. Kirkwood and Shumaker (20) have shown that these charge fluctuations result in a significant attractive force between isoionic protein molecules if the ionic strength is low. Using the solution theory of Kirkwood and Buff (98), Kirkwood and Shumaker have calculated the effect of charge fluctuations on the chemical potential of the protein. Setting the protein as component 2 and water as component 1, according to Kirkwood and Buff, at low protein concentration

$$\left(\frac{\partial \mu_2^{(e)}}{\partial C_2}\right)_{T,p} = \frac{RT}{M_2} (G_{21} - G_{22}) \tag{77}$$

$$G_{22} = 4\pi \int_0^\infty R^2 [g_{11}(R) - 1] \, dR$$

$$G_{21} = 4\pi \int_0^\infty R^2 [g_{21}(R) - 1] \, dR$$

where C_2 is the protein concentration in g/ml, M_2 is its molecular weight and $g_{22}(R)$ and $g_{21}(R)$ are the radial distribution functions of a pair of protein molecules and a protein and solvent molecule, respectively; R is the distance between the molecules in each pair. These are related to the potentials of average force $W_{ij}(R)$, by

$$g_{ij}(R) = e^{-W_{ij}(R)/kT} \tag{78}$$

$W_{22}(R)$ is calculated from the potential, V, in fixed orientation and proton configuration by (99)

$$W(R) = \langle V_{\mathrm{Av}} \rangle = \frac{1}{2kT} [\langle V^2 \rangle_{\mathrm{Av}} - \langle V_{\mathrm{Av}} \rangle^2] \tag{79}$$

Using a screened Coulombic potential for V_{Av}, at the point where $\bar{Z}_p = 0$, we obtain

$$W_{22}(R) = -\frac{\langle Z_2^2 \rangle^2_{\mathrm{Av}} \epsilon^4}{2kTD^2R^2} \frac{e^{-\kappa(R-a)}}{1 + \kappa a} \tag{80}$$

where R is the distance between the pair of protein molecules and all other symbols have their previous meaning. This gives finally:

$$\frac{1}{RT}\left(\frac{\partial \mu_2^{(e)}}{\partial C_2}\right)_{T,p} = \frac{7\pi N b^3}{6M_2} - \frac{\pi N \epsilon^4 \langle Z_2^2 \rangle^2_{\mathrm{Av}}}{M_2(DkT)^2\kappa(1 + \kappa a)^2} \tag{81}$$

$$\kappa^2 = \frac{4\pi N \epsilon^2}{DkT}\left[\frac{\displaystyle\sum_j m_j Z_j^2}{1000} + \frac{\langle Z_2^2 \rangle_{\mathrm{Av}} C}{M_2}\right]$$

where m_j is the molar concentration of supporting electrolyte ions of species j, and Z_j is their valence. Binomial expansion of $(1 + \kappa a)^{-2}$ in Eq. 81 gives:

$$\frac{C_2}{RT}\left(\frac{\partial \mu_2^{(e)}}{\partial C_2}\right)_{T,p} = -\frac{\pi^{1/2} N^{1/2} \epsilon^3 \langle Z_2^2 \rangle^{3/2}_{\mathrm{Av}}}{2(DkT)^{3/2}M_2^{1/2}} C_2^{1/2} + 0(C) \tag{82}$$

Examination of Eq. 81 and 82 shows that the effect of charge fluctuations, when the net average charge of the protein is zero, is a net attraction which in the limit is linear in the square root of protein concentration. Such a concentration dependence of light scattering has been observed with serum albumin (*100*) and conalbumin (*24*). When the net average charge on the protein is not zero, Eqs. 81 and 82 become more complicated, now including terms which involve $\bar{Z}_p$. This case has been treated by Timasheff and Coleman (*101*).

When the isoionic point of the protein is not at pH 7.0, as has been pointed out, electroneutrality requires that the net average charge increase as protein concentration decreases. This effect makes a repulsive contribution to the intermolecular interactions. Kirkwood and Timasheff (*102*) have shown that the effect of the progressive ionization of the isoionic protein as its concentration decreases on its chemical potential is:

$$\frac{1}{RT}\left(\frac{\partial \mu_2^{(e)}}{\partial m_2}\right)_{T,p}^{(pr.\ ion.)} = \frac{\bar{Z}_2^2}{[H^+]} \frac{1}{1 + \dfrac{K_w}{[H^+]^2} + m_2 \dfrac{d\bar{Z}_2}{d[H^+]}} \tag{83}$$

where K_w is the dissociation constant of water and m_2 is the protein concentration in moles/liter at low protein concentration. In the case of bovine serum albumin which has an isoelectric point close to pH 5, this effect may be quite large. In Fig. 16 its magnitude is shown in light-

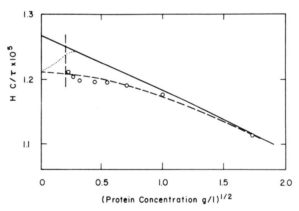

Fig. 16. Effect of progressive ionization on the light scattering of isoionic bovine serum albumin in $1 \times 10^{-5} M$ HCl (region below 0.3% protein) (*25*). Solid line: least-squares cure of the experimental points; ○ = experimental points from which the ionization term had been subtracted. Dashed line: curve fitting best the corrected points (extrapolated to zero protein concentration). Dotted line: curve obtained by adding calculated values of ionization term to the dashed line. (The vertical dashed line represents the lower limit of experimental data, 0.005%.) In a two component system, the light-scattering parameter

$$H\frac{C}{\tau} = \frac{1}{M_2}\left[1 + \frac{C_2}{RT}\left(\frac{\partial \mu_2^{(e)}}{\partial C_2}\right)_{T,p}\right]$$

scattering experiments on serum albumin in the presence of $1 \times 10^{-5} M$ HCl.

3. Long-Range Order

Under conditions at which the protein carries a high net charge and is in the presence of little supporting electrolyte, very strong repulsive forces set in. This situation has been examined theoretically by Kirkwood and Mazur (103,104) and by Fournet (105–107). Using the Born and Green (108) integral equation, Kirkwood and Mazur have calculated the radial distribution function (103) for a pair of charged spherical particles by using the Verwey-Overbeek potential (Eq. 31). Their conclusion was that in such a system long-range order should set in and, at sufficiently high charge and low concentration of small electrolyte, should result in gelation of the solution because of repulsive forces alone (104).

Applying the Fournet treatment, Doty and Steiner (109,110) examined this case experimentally in light-scattering experiments. According to Zernicke and Prins, the attenuation of scattering per particle because of external interference due to long-range ordering in solution is (111,112)

$$
\frac{I_\theta}{I_0} = 1 - \frac{4\pi}{v_1} \int_0^\infty [1 - g(R)] \frac{\sin hR}{hR} R^2 \, dR
$$
$$
h = \frac{4\pi \sin(\theta/2)}{\lambda'}
$$

(84)

where I_θ and I_0 are the light or X-ray intensities of the scattered beam at angle θ and at zero angle, respectively, v_1 is the solution volume per particle, λ' is the wavelength of the light in the medium, and θ is the angle between the incident and scattered beams. At the usual angle of observation, $\theta = 90°$, this effect should result in a very large deviation from ideality, i.e., a large positive, concentration-dependent value of $(\partial \mu_2^{(e)} / \partial C_2)_{T,p}$. The results of light-scattering experiments on bovine serum albumin in water adjusted to various pH's with hydrochloric acid are shown in Fig. 17 (110). It is evident that repulsion between highly charged protein molecules results in a dominant concentration-dependent term. With an increase of supporting electrolyte and a decrease in protein charge, this effect reduces essentially to the virial coefficient treatment of Stigter and Hill (95,113).

4. Interactions with Small Ions and Multicomponent Effects

Although these topics will not be discussed in detail in the present chapter, they must nevertheless be mentioned. Proteins bind ions other than

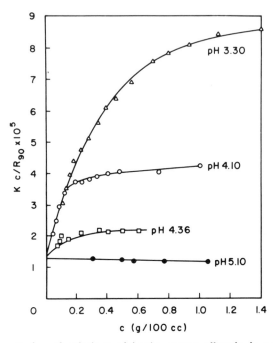

Fig. 17. Light scattering of solutions of bovine serum albumin in water adjusted to the indicated pH values with hydrochloric acid (*110*).

hydrogen; this may result in strong activity coefficient effects. Using the Donnan model, Scatchard (*114*) has derived equations for the chemical potential and osmotic pressure, while Edsall et al. (*115*) have extended this treatment to light scattering and Johnson, Kraus, and Scatchard (*116*) have applied it to sedimentation equilibrium. This work has been the subject of a detailed review recently (*117*); therefore, it will be presented only in outline form. Calling the protein component 2 and the salt component 3, Scatchard defined the components in such a way that they are electrically neutral. This requires adding $\bar{Z}_2/2$ moles of the diffusible ion with a sign of charge opposite to that of the protein and removing $\bar{Z}_2/2$ moles of the ion with the same sign of charge as the protein when one mole of protein is added to the system. As a result, addition of the protein leaves the system electrically neutral and involves no net addition of diffusible ions. The resulting effect on the activity of the protein is:

$$\frac{\mu_2}{RT} = \ln m_2 + \sum_i \nu_{2i} \ln m_i + \ln \gamma_2 + \frac{\mu_2^0(T, p)}{RT} \qquad (85)$$

and for a one-one supporting electrolyte

$$\frac{1}{RT}\left(\frac{\partial \mu_2}{\partial m_2}\right) = \frac{1}{m_2} + \sum_i \frac{\nu_{2i}^2}{m_i} + \beta_{22}$$

$$= \frac{1}{m_2} + \frac{\bar{Z}_2^2}{2m_3\delta} + \beta_{22}$$

$$\beta_{ij} = \frac{\partial \ln \gamma_i}{\partial m_j} = \frac{1}{RT}\left(\frac{\partial \mu_i^{(e)}}{\partial m_j}\right)$$

$$\delta = 1 - (\bar{Z}_2 m_2/2m_3)^2 \tag{86}$$

where ν_{2i} is the number of moles of ionic species i added to the system per protein molecule and m_i is the molal concentration of that species in the system. This result can be applied to treating nonideality in both sedimentation equilibrium and light scattering (and also small angle X-ray scattering). For the case of a one-one supporting electrolyte, the appropriate equations for dilute protein solutions are

$$\frac{1}{M_{2,\text{app}}} = \frac{1}{M_2(1+D)^2}\left[1 + \frac{1000C_2}{M_2}\left(\frac{\bar{Z}_2^2}{2m_3\delta} + \beta_{22} - \frac{\beta_{23} - \dfrac{1000\bar{Z}_2^2 C_2}{2M_2 m_3^2\delta}}{\dfrac{2}{m_3\delta} + \beta_{33}}\right)\right] \tag{87}$$

For light scattering,†

$$\frac{1}{M_{2,\text{app}}} = \frac{H'C_2}{\tau}\left(\frac{\partial n}{\partial C_2}\right)_{T,p,m_3}$$

and $\tag{87a}$

$$1 + D = 1 + \frac{(1 - C_3\bar{V}_3)_{m_2}}{(1 - C_2\bar{V}_2)_{m_3}}\frac{(\partial n/\partial C_3)_{T,p,m_2}}{(\partial n/\partial C_2)_{T,p,m_3}}\left(\frac{\partial g_3}{\partial g_2}\right)_{T,p,\mu_3}$$

For sedimentation equilibrium

$$\frac{1}{M_{2,\text{app}}} = \frac{2RT}{\omega^2(1 - \bar{V}_2\rho)_{m_3}}\frac{d \ln C_2}{dr^2}$$

and $\tag{87b}$

$$(1 + D)^2 = 1 + \frac{(1 - \bar{V}_3\rho)_{m_2}}{(1 - \bar{V}_2\rho)_{m_3}}\left(\frac{\partial g_3}{\partial g_2}\right)_{T,p,\mu_3}$$

$$\left(\frac{\partial g_3}{\partial g_2}\right)_{T,p,\mu_3} = -\frac{M_3}{M_2}\frac{\beta_{32}}{\dfrac{2}{m_3} + \beta_{33}}$$

† Essentially the same equation applies to small angle X-ray scattering; it is only necessary to replace the refractive index increment by the electron density increment.

In these equations, M_2 is the true molecular weight of the protein, $\bar{Z}_2$ is its average charge, C is concentration in grams/milliliter, $(\bar{V}_i)_{m_j}$ is the partial specific volume of component i measured in the usual way, i.e., keeping the molality of component j equal in the protein solution and the reference standard, ρ is the density of the solution, R is the gas constant, T is the thermodynamic temperature, ω^2 is the angular acceleration, r is the radial distance from the center of rotation, τ is the excess turbidity of the protein solution over that of the solvent, H' is an optical constant, $(\partial n/\partial C_i)_{T,p,m_j}$ is the refractive index increment of the solution on addition of component i, keeping the molal concentration of component j constant, and g_i is the concentration of component i in grams/gram of principal solvent, i.e., water. $(\partial g_3/\partial g_2)_{T,p,\mu_3}$ represents the preferential interaction between the protein and component 3, i.e., the excess of component 3 in the immediate domain of component 2 over its concentration in the bulk solvent. It should be noted that this term does not extrapolate out at zero protein concentration. Thus, in a three (or more)-component system, the extrapolated value of $M_{2,\text{app}}$ is a product of the true protein molecular weight and a measure of the interaction between protein and solvent components. The interaction term, $(\partial g_3/\partial g_2)_{T,p,\mu_3}$, may be either positive or negative; in the first case it means that component 3 (in our case supporting electrolyte) is present in excess in the immediate vicinity of the protein, in the second case there is a deficiency of this component near the protein, i.e., water is favored in the vicinity of the macromolecule. Thus, from such measurements it is possible to determine the preferential binding of small components to the protein as well as the changes in activity coefficient of the protein induced by the addition of various other components to the system (118). Typical results obtained with serum albumin, shown on Fig. 18, indicate the magnitude of the salt binding effect on the activity coefficient of the protein in the case of dilute salts.

The effect of protein charge on sedimentation equilibrium has been treated by Johnson, Kraus, and Scatchard (116). They define the electroneutral components as PX_z and BX, where P is protein of charge Z, X is the gegenion, and BX is the supporting one-one electrolyte. In a two-component system, PX_z and water, the equilibrium sedimentation equation is:

$$\frac{2RT}{(1 - \bar{V}\rho)\omega^2} \frac{d \ln a_{\pm PX_z}}{d(r^2)} = \frac{M_{PX_z}}{Z + 1} \tag{88}$$

where $\bar{V}$ is the partial specific volume of the protein, ρ is the solution density, r is the radial distance from center of rotation, and $a_{\pm PX}$ is the activity of the electroneutral species; we see that in a two-component system, at extrapolation to infinite dilution, equilibrium sedimentation gives not the

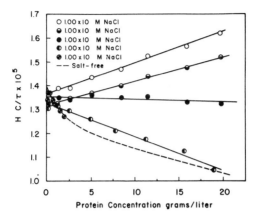

Fig. 18. Light scattering of isoionic bovine serum mercaptalbumin in various concentrations of NaCl (*25*).

molecular weight of the electroneutral species, M_{PX_z}, but the number-average molecular weight of one protein ion, P, and the z gegenions, X. This equation has been verified on silico-tungstic acid (*119*) and the results have been compared with those of light scattering, which should give identical results, with good agreement between the two techniques (*119,120*).

The three-component system was treated within the following restrictions:

1. The polymer dissociates: $PX_z \rightleftharpoons P^{+z} + zX^-$; the supporting electrolyte is: $BX \rightleftharpoons B^+ + X^-$.
2. The polymer is monodisperse.
3. The partial specific volumes, $\bar{V}$, of PX_z and BX are constant.
4. The density of the solution, ρ, is constant.
5. The activity coefficient products of all species are constant; thus, $d \ln \gamma_P \gamma_X^z = (Z+1) \, d \ln \gamma_{\pm,PX_z} = 0$, and $d \ln \gamma_B \gamma_X = 0$.

The equilibrium distribution of the two solutes is given by the simultaneous solution of the three equations:

$$d \ln a_{PX_z} = d \ln m_P m_X^z = \frac{M_2(1 - \bar{V}_2 \rho)\omega^2}{2RT} d(r^2)$$

$$d \ln a_{BX} = d \ln m_B m_X = \frac{M_3(1 - \bar{V}_3 \rho)\omega^2}{2RT} d(r^2)$$

$$m_X = z m_P + m_B \tag{89}$$

For the case where the polymer concentration is vanishingly small (i.e., for extrapolation to zero protein concentration), so that we may set $m_X \simeq m_B$ and $d \ln m_X = d \ln m_B$, this gives

$$M_{app} = \frac{2RT}{(1 - \bar{V}_2\rho)\omega^2} \frac{d \ln a_2}{d(r^2)} = M_2 - \frac{(1 - \bar{V}_3\rho)}{(1 - \bar{V}_2\rho)} \frac{Z_2}{2} M_3 \qquad (90)$$

where Z_2 is the net charge on the protein, M_3 is the molecular weight of
the salt BX, M_2 that of protein PX_z, and all other symbols have their usual
meaning. We see that if $M_3 << M_2$ and the charge is not extreme, no
very important errors should be introduced.

VI. EFFECT OF PROTEIN CHARGE ON TRANSPORT PROCESSES

So far we have discussed the effects which the polyelectrolyte properties
of proteins have on the equilibrium properties of these macromolecules.
Some consideration should be given also to the manifestations of the protein electric charges in nonequilibrium phenomena, in particular to transport processes, such as electrophoresis, sedimentation, and viscosity.

A. Electrophoresis

The theory of electrophoresis was treated first by Smoluchowski in 1903
(121). Since then it has been greatly extended and refined (122–130).

When a charged protein molecule is subjected to a dc electric field, the
particle starts moving toward the proper electrode and a stationary state is
rapidly attained. The forces acting on the particle in electrophoretic motion are indicated on Fig. 19. They are:

1. The force exerted by the dc field, E, on the net charge of the particle; it is equal to QE, where $Q = Z_2\epsilon$.

2. The retarding frictional force, equal to $-f_cU$, where U is the electrophoretic velocity and f_c is the frictional coefficient. For a spherical particle, $f_c = 6\pi\eta b'$.

3. The force exerted by the dc field on the ions of the ionic atmosphere.

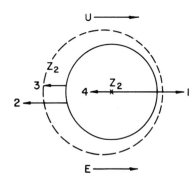

Fig. 19. Forces acting on a charged particle in electrophoresis (129).

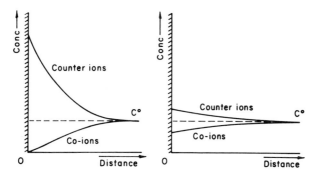

Fig. 20.   Distribution of ions in the field of an electrically charged particle.   Left: for high potentials $(F\psi/RT > 1)$.   Right: for low potentials $(F\psi/RT \ll 1)$.   F is the Faraday (130a).

Since these ions are distributed unequally next to the protein molecule, as shown in Fig. 20, the ionic atmosphere has a net charge opposite in sign to the protein (129).   This force causes a retardation.

4. The deformation of the ionic distribution in the vicinity of the protein due to the motion of the latter away from the center of its ionic atmosphere.

In the stationary state, the sum of all forces acting on the particle is zero. Neglecting the third and fourth effects,

$$u = \frac{U}{E} = \frac{Q}{6\pi\eta b'} \tag{91}$$

where u is the electrophoretic mobility or particle velocity per unit applied field, usually given in the units of $cm^2/V\text{-}sec$ (131).   The question arises as to the values of Q and b'.   When a particle moves, a thin layer of liquid usually remains fixed to it.   The boundary between this "fixed" liquid and the "free" liquid is called the surface of shear.   The charge Q, therefore, must be taken at that surface, as must also the radius b'.   This charge is related to the potential at the surface of shear, ζ, by

$$Q = D\zeta b' \left[\frac{1 + \kappa(b' + a_i)}{1 + \kappa a_i} \right] \tag{92}$$

where a_i is the radius of the ions of the supporting electrolyte and D is the dielectric constant of the medium.

By taking the third effect into account, Henry (124) derived the relation:

$$u = \frac{D\zeta}{6\pi\eta} f(\kappa b') \tag{93}$$

where $f(\kappa b')$ is a complicated function of the radius and ionic strength which varies between 1 when $\kappa b' \to 0$ and $\frac{3}{2}$ when $\kappa b' \to \infty$.

The last (fourth contribution), the ionic atmosphere relaxation effect, has been treated by Booth (126) and Overbeek (127). Using a Gouy–Chapman model of the ionic atmosphere and series expansion of the Poisson–Boltzmann equation of the type given by Gronwall et al. (132,133), the two treatments result in an equation of the form:

$$u = \frac{D\zeta}{6\pi\eta}\,[f(\kappa b') + C_2\zeta + C_3\zeta^2 + C_4\zeta^3 + \cdots]\qquad(94)$$

This is seen to differ from Henry's equation by the terms in $C_n\zeta^{n-1}$. These terms are complicated functions of $\kappa b'$ and the valences and mobilities of the small ions. They have been evaluated up to C_4. With the availability of electronic computers, the electrophoretic mobility as a function of charge, ionic strength, and particle radius could be calculated more exactly (130,134,135). The results of the computation are given in Fig. 21; they

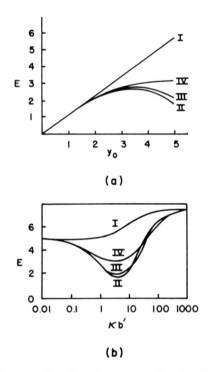

(a)

(b)

Fig. 21. (a) Plot of E as a function of y_0 at $\kappa b' = 5$. (b) Plot of E as a function of $\kappa b'$ at $y_0 = 5$. I: Henry (Eq. 93). II: Overbeek (Eq. 94). III: Booth (Eq. 94). IV: Wiersema, numerical computation. These calculations are for $Z_+ = Z_- = 1$; $m_+ = m_- = 0.184$; $m \equiv (NDkT/6\pi\eta)(Z/\lambda_0)$, where λ_0 is the limiting equivalent conductance of the small ion (128).

are presented in terms of the relation between the dimensionless parameter E and the quantity y_0, where

$$E = \frac{6\pi\eta\epsilon}{DkT}\,u; \qquad y_0 = \frac{\epsilon\zeta}{kT} \tag{95}$$

and also as a function of $\kappa b'$. Several conclusions are evident: (1) The relaxation correction is large and increases strongly with increasing ζ-potential; (2) this effect is largest at moderate values of $\kappa b'$; (3) at low ζ-potentials Eq. 94 gives correct results, it overestimates the effect of high values of ζ, and at moderate $\kappa b'$ values a maximum will appear in the u vs ζ curve.

The fixed layer of liquid on the particle may be expected to increase the viscosity of the solution when an electric field is applied. This is called the viscoelectric effect. From a comparison of experimental data with theory (136), Stigter (137) came to the conclusion that the surface of shear comes within 1 Å of the hydrated surface of sodium dodecyl sulfate micelles. Using a treatment similar to that of electrophoresis, Booth (138) has analyzed the decrease in sedimentation velocity due to the electric field resulting from the charge of the particle. For the case of suspensions, such as protein solutions, he was able to obtain results only for the limiting cases of very large and very small thicknesses of the double layer. The calculations show that in practical cases the effect might reduce the velocity by a few percent.

B. Electroviscous Effect

When a charged particle, such as a protein, is suspended in an electrolyte, the double layer around the particle may be expected to increase the viscosity of the suspension (139). Booth (140) examined this problem in detail and showed that the electroviscous effect, η_{el}, is equal to:

$$\eta_{el} = \eta_0 \left[\frac{5v}{2V} \left\{ 1 + q^* \left(\frac{\epsilon^2 Z}{DbkT} \right)^2 F(\kappa b) \right\} \right] \tag{96}$$

$$q^* = DkT \sum_{i=1}^{s} n_i z_i^2 \omega_i^{-1} / \eta_0 \epsilon^2 \sum n_i z_i^2$$

where η_0 is the viscosity of the solvent, v is the volume occupied by the particles, V is the total volume of the solution, ϵ is the electronic charge, Z is the net charge of the particle, D is the dielectric constant of the medium, b is the radius of the particle, k is Boltzmann's constant, T is the thermodynamic temperature, n_i is the number of ions of species i, z_i is their valence, ω_i is the ionic mobility, and $F(\kappa b)$ is a complicated function written out by Booth. Calculations have shown that η_{el} increases with increase in charge and decrease in ionic strength. At the usual working ionic strengths (0.01–0.3), this effect is found to be very small.

REFERENCES

1. C. Tanford, *Physical Chemistry of Macromolecules,* Wiley, New York, 1961, p. 394.
2. M. E. Noelken, C. A. Nelson, C. E. Buckley, III, and C. Tanford, *J. Biol. Chem.,* **240,** 218 (1965).
3. J. C. Kendrew, H. C. Watson, B. E. Strandberg, R. E. Dickerson, D. C. Phillips, and V. C. Shore, *Nature,* **190,** 666 (1961).
4. D. Phillips, *Sci. Amer.,* **215,** 78 (1966).
5. H. Wyckoff, K. Hardman, H. Allewell, T. Inogami, D. Tsernoglan, L. Johnson, and F. M. Richards, *J. Biol. Chem.,* **242,** 3749 (1967).
6. B. W. Matthews, P. B. Sigler, R. Henderson, and D. M. Blow, *Nature,* **214,** 652 (1967).
7. For a discussion of this problem, see R. Lumry and R. Biltonen, in *Structure and Stability of Biological Macromolecules* (S. N. Timasheff and G. D. Fasman, eds.), Dekker, New York, 1969, Chap. 2.
8. M. A. Naughton and H. M. Dintzis, *Proc. Nat. Acad. Sci. U.S.,* **48,** 1822, (1962).
9. C. Tanford, *J. Amer. Chem. Soc.,* **84,** 4240 (1962).
10. C. Tanford and J. G. Kirkwood, *J. Amer. Chem. Soc.,* **79,** 5333 (1957).
11. C. Tanford, *Advan. Protein Chem.,* **17,** 69 (1962).
12. Y. Nozaki and C. Tanford, *J. Biol. Chem.,* **242,** 4731 (1967).
13. Y. Nozaki and C. Tanford, *J. Amer. Chem. Soc.,* **89,** 736 (1967).
14. E. J. Cohn and J. T. Edsall, *Proteins, Amino Acids and Peptides,* Reinhold, New York, 1943, Chap. 20.
15. J. T. Edsall and J. Wyman, *Biophysical Chemistry,* Vol. 1, Academic, New York, 1958, Chap. 9.
16. S. A. Rice and M. Nagasawa, *Polyelectrolyte Solutions,* Academic, New York, 1961.
17. J. Steinhardt and S. Beychok, *The Proteins* (H. Neurath, ed.), Vol. 2, Academic, New York, 1964, Chap. 8.
18. K. V. Linderstrøm-Lang, *C. R. Trav. Lab. Carlsberg,* **15,** 1 (1924).
19. J. G. Kirkwood and J. B. Shumaker, *Proc. Nat. Acad. Sci. U. S.,* **38,** 855 (1952).
20. J. G. Kirkwood and J. B. Shumaker, *Proc. Nat. Acad. Sci. U. S.,* **38,** 863 (1952).
21. S. N. Timasheff and B. D. Coleman, *Arch. Biochem. Biophys.,* **87,** 63, (1960).
22. G. Scatchard, in *Proteins, Amino Acids and Peptides* (E. J. Cohn and J. T. Edsall, eds.), Reinhold, New York, 1943, Chap. 2.
23. S. Lowey, *Arch. Biochem. Biophys.,* **64,** 111 (1956).
24. S. N. Timasheff and I. Tinoco, Jr., *Arch. Biochem. Biophys.,* **66,** 427 (1957).
25. S. N. Timasheff, H. M. Dintzis, J. G. Kirkwood, and B. D. Coleman, *J. Amer. Chem. Soc.,* **79,** 782 (1957).
26. K. V. Linderstrøm-Lang and J. A. Schellman, in *The Enzymes* (P. D. Boyer, H. Lardy and K. Myrbäck, eds.), Academic, New York, 1959, Chap. 10.
27. J. A. Schellman, *J. Phys. Chem.,* **62,** 1485 (1958).
28. A. Hirdt and S. O. Nielsen, *Advan. Protein Chem.,* **21,** 288 (1966).

29. P. Flory and W. Miller, *J. Mol. Biol.,* 15, 284 (1966).
30. W. Miller, D. Brant, and P. Flory, *J. Mol. Biol.,* 23, 67 (1967).
31. C. Tanford, *Physical Chemistry of Macromolecules,* Wiley, New York, 1961, Chap. 6.
32. T. Svedberg and K. O. Pedersen, *The Ultracentrifuge,* Oxford University Press, Oxford, 1940.
33. R. Townend, R. J. Winterbottom, and S. N. Timasheff, *J. Amer. Chem. Soc.,* 82, 3161 (1960).
34. R. Townend, L. Weinberger, and S. N. Timasheff, *J. Amer. Chem. Soc.,* 82, 3175 (1960).
35. S. N. Timasheff and R. Townend, *J. Amer. Chem. Soc.,* 83, 464 (1961).
36. J. Witz, S. N. Timasheff, and V. Luzzati, *J. Amer. Chem. Soc.,* 86, 168 (1964).
37. R. Townend and S. N. Timasheff, in preparation.
38. D. W. Green and R. Aschaffenburg, *J. Mol. Biol.,* 1, 54 (1961).
39. S. N. Timasheff and R. Townend, *Nature,* 203, 517 (1964).
40. M. Nagasawa and A. Holtzer, *J. Amer. Chem. Soc.,* 86, 531 (1964).
41. R. A. Robinson and R. H. Stokes, *Electrolyte Solutions,* Academic, New York, 1955.
42. C. Tanford, *J. Amer. Chem. Soc.,* 79, 5340 (1957).
43. C. Tanford, *J. Amer. Chem. Soc.,* 79, 5348 (1957).
44. T. L. Hill, *J. Amer. Chem. Soc.,* 78, 5527 (1956).
45. T. L. Hill, *J. Amer. Chem. Soc.,* 78, 1577 (1956).
46. C. Tanford, *Physical Chemistry of Macromolecules,* Wiley, New York, 1961, Chap. 7.
47. T. L. Hill, *J. Chem. Phys.,* 12, 147 (1944).
48. K. V. Linderstrøm-Lang, *C. R. Trav. Lab. Carlsberg,* 28, 281 (1953).
49. T. L. Hill, *Arch. Biochem. Biophys.,* 57, 229 (1955).
50. C. Tanford, *J. Phys. Chem.,* 59, 788 (1955).
51. V. Luzzati, J. Witz, and A. Nicolaieff, *J. Mol. Biol.,* 3, 367, 379 (1961).
52. A. Katchalsky, N. Shairt, and H. Eisenberg, *J. Polym. Sci.,* 13, 69 (1954).
53. C. Tanford and J. Epstein, *J. Amer. Chem. Soc.,* 76, 2163 (1954).
54. E. J. Verwey and J. Th. G. Overbeek, *Theory of the Stability of Lyophobic Colloids,* Elsevier, Amsterdam, 1948.
55. C. Tanford, *Physical Chemistry of Macromolecules,* Wiley, New York, 1961, Chap. 8.
56. G. Scatchard and E. S. Black, *J. Phys. Colloid Chem.,* 53, 88 (1949).
57. M. Laskowski, Jr., and H. A. Scheraga, *J. Am. Chem. Soc.,* 76, 6305 (1954).
58. C. Tanford, *J. Amer. Chem. Soc.,* 83, 1628 (1961).
59. J. Steinhardt, R. Ona, and S. Beychok, *Biochemistry,* 1, 29 (1962).
60. L. M. Riddiford, *J. Biol. Chem.,* 239, 1079 (1964).
61. L. M. Riddiford, R. H. Stellwagen, S. Mehta, and J. T. Edsall, *J. Biol. Chem.,* 240, 3305 (1965).
62. M. J. Gorbunoff, *Biochemistry,* 6, 1606 (1967).
63. M. J. Gorbunoff, *Biochemistry,* 7, 2547 (1968).
63a. M. J. Gorbunoff, *Biochemistry,* 8, 2591 (1969).
64. S. N. Timasheff and M. J. Gorbunoff, *Ann. Rev. Biochem.,* 36, 13 (1967).
65. C. Tanford, L. G. Bunville, and Y. Nozaki, *J. Amer. Chem. Soc.,* 81, 4032 (1959).
66. C. Tanford and Y. Nozaki, *J. Biol. Chem.,* 234, 2874 (1959).
67. H. Susi, T. Zell, and S. N. Timasheff, *Arch. Biochem. Biophys.,* 85, 437 (1959).

68. J. J. Basch and S. N. Timasheff, *Arch. Biochem. Biophys.,* 118, 37 (1967).
69. A C. Ghose, S. Chandhuri, and A. Sen, *Arch. Biochem. Biophys.,* 126, 232 (1968).
70. S. N. Timasheff, L. Mescanti, J. J. Basch, and R. Townend, *J. Biol. Chem.,* 241, 2496 (1966).
71. E. E. Rickli, S. A. A. Ghazanfar, B. H. Gibbons, and J. T. Edsall, *J. Biol. Chem.,* 239, 1065 (1964).
72. S. Beychok, J. McD. Armstrong, C. Lindblow, and J. T. Edsall, *J. Biol. Chem.,* 241, 5150 (1966).
73. C. Tanford and V. G. Taggart, *J. Amer. Chem. Soc.,* 83, 1634 (1961).
74. C. F. Jacobsen and K. V. Linderstrøm-Lang, *Nature,* 164, 411 (1949).
75. G. E. Perlmann, A. Oplatka, and A. Katchalsky, *J. Biol. Chem.,* 242, 5163 (1967).
76. T. M. Birshtein and O. B. Ptitsyn, *Conformations of Macromolecules,* Wiley (Interscience), New York, 1966, Chap. 10.
77. J. Lebowitz and M. Laskowski, Jr., *Biochemistry,* 1, 1044 (1962).
77a. M. Laskowski, Jr., private communication.
78. L. W. Nichol and D. J. Winzor, *Biochim. Biophys. Acta,* 94, 591 (1965).
78a. G. A. Gilbert, private communication.
79. J. M. Sturtevant, M. Laskowski, Jr., T. H. Donnelly, and H. A. Scheraga, *J. Amer. Chem. Soc.,* 77, 6168 (1955).
80. G. F. Endres, and S. Ehrenpreis, and H. A. Scheraga, *Biochemistry,* 5, 1561 (1966).
81. G. F. Endres and H. A. Scheraga, *Biochemistry,* 5, 1568 (1966).
82. H. A. Scheraga, *Protein Structure,* Academic, New York, 1961, Chap. 2.
83. H. A. Scheraga, *The Proteins* Vol. 1 (H. Neurath, ed.), Academic, New York, 1963, Chap. 6.
84. T. F. Kumosinski and S. N. Timasheff, *J. Amer. Chem. Soc.,* 88, 5635, (1966).
85. S. N. Timasheff and R. Townend, in *Proceedings of the 16th Colloquium on Protides of Biological Fluids,* Vol. 16 (H. Peeters, ed.), Pergamon, New York, 1969, p. 33.
86. S. N. Timasheff, *Proteins and Their Reactions* (H. W. Schultz and A. F. Anglemier, eds.), Avi, Westport, Conn., 1964, p. 179.
87. M. R. Bruzzesi, E. Chiancone, and E. Antonini, *Biochemistry,* 4, 1796 (1965).
87a. S. N. Timasheff, *Arch. Biochem. Biophys.,* 132, 165 (1969).
88. R. Egan, H. O. Michel, R. Schlueter, and B. J. Jandorf, *Arch. Biochem. Biophys.,* 66, 366 (1957).
89. M. A. Marini and C. Wunsch, *Biochemistry,* 2, 1454 (1963).
90. G. A. Gilbert, *Discussions Faraday Soc.,* No. 2, 68 (1955).
91. G. A. Gilbert, *Proc. Roy. Soc., Ser. A* (London), 250, 377 (1959).
91a. J. J. Birktoft, B. W. Matthews, and D. M. Blow, *Biochem. Biophys. Research Communs.,* 36, 131 (1969).
91b. K. C. Aune and S. N. Timasheff, in preparation.
92. S. N. Timasheff, *Biopolymers,* 4, 107 (1966).
93. T. L. Hill, *J. Amer. Chem. Soc.,* 78, 3330 (1956).
94. W. G. McMillan and J. E. Mayer, *J. Chem. Phys.,* 13, 276 (1945).
95. D. Stigter and T. L. Hill, *J. Phys. Chem.,* 63, 551 (1959).
96. N. E. Hoskins, *Trans. Faraday Soc.,* 49, 1471 (1953).
97. D. Stigter, *J. Phys. Chem.,* 64, 838 (1960).
98. J. G. Kirkwood and F. P. Buff, *J. Chem. Phys.,* 19, 774 (1951).
99. J. G. Kirkwood, *Discussions Faraday Soc.,* No. 20, 78 (1955).

100. S. N. Timasheff, H. M. Dintzis, J. G. Kirkwood, and B. D. Coleman, *J. Amer. Chem. Soc.,* **79**, 782 (1957).

101. S. N. Timasheff and B. D. Coleman, *Arch. Biochem. Biophys.,* **87**, 63 (1960).

102. J. G. Kirkwood and S. N. Timasheff, *Arch. Biochem. Biophys.,* **65**, 50 (1956).

103. J. G. Kirkwood and J. Mazur, *J. Polym. Sci.,* **9**, 519 (1952).

104. J. G. Kirkwood and J. Mazur, *Comptes. Rendus. 2° Réunion Chim. Phys.,* Paris, 1952, p. 143.

105. G. Fournet, *C. R. Acad. Sci., Paris,* **228**, 1421, 1801 (1949).

106. G. Fournet, *Acta Crystallogr.,* **4**, 293 (1951).

107. G. Fournet, *Bull. Soc. Fr. Mineral. Cristallogr.,* **74**, 37 (1951).

108. M. Born and H. S. Green, *Proc. Roy. Soc., Ser. A* (London), **188**, 10 (1946).

109. P. Doty and R. F. Steiner, *J. Chem. Phys.,* **17**, 743 (1949).

110. P. Doty and R. F. Steiner, *J. Chem. Phys.,* **20**, 85 (1952).

111. F. Zernicke and P. A. Prins, *Z. Phys.,* **41**, 184 (1927).

112. P. Debye and H. Menke, *Phys. Z.,* **31**, 797 (1930).

113. D. Stigter, *J. Amer. Chem. Soc.,* **64**, 842 (1960).

114. G. Scatchard, *J. Amer. Chem. Soc.,* **68**, 2315 (1946).

115. J. T. Edsall, H. Edelhoch, R. Lontie, and P. R. Morrison, *J. Amer. Chem. Soc.,* **72**, 4641 (1950).

116. J. S. Johnson, K. A. Kraus, and G. Scatchard, *J. Phys. Chem.,* **58**, 1034 (1954).

117. E. F. Casassa and H. Eisenberg, *Advan. Protein Chem.,* **19**, 287 (1964).

118. For a detailed review, see S. N. Timasheff and R. Townend, in *Principles and Methods of Protein Chemistry* (S. Leach, ed.), Academic, New York, 1970, in press.

119. J. S. Johnson, K. A. Kraus, and G. Scatchard, *J. Phys. Chem.,* **64**, 1867 (1960).

120. S. N. Timasheff and L. D. Cerankowski, in preparation.

121. M. von Smoluchowski, *Bull. Acad. Sci. Cracovie,* **1903**, 182.

122. P. Debye and E. Hückel, *Phys. Z.,* **25**, 49 (1924).

123. E. Hückel, *Phys. Z.,* **25**, 204 (1924).

124. D. C. Henry, *Proc. Roy. Soc., Ser. A* (London), **133**, 106 (1931).

125. M. H. Gorin, *J. Chem. Phys.,* **7**, 405 (1939).

126. F. Booth, *Nature,* **161**, 83 (1948).

127. J. Th. G. Overbeek, *Advan. Colloid Sci.,* **3**, 97 (1950).

128. J. Th. G. Overbeek and P. H. Wiersema, in *Electrophoresis,* Vol. 2 (M. Bier, ed.), Academic, New York, 1967, p. 1.

129. J. Th. G. Overbeek, *Progr. Biophys. Biophys. Chem.,* **6**, 58 (1956).

130. P. H. Wiersema, Ph.D. thesis, Univ. of Utrecht, 1964.

131. R. A. Brown and S. N. Timasheff, in *Electrophoresis,* Vol. 1 (M. Bier, ed.), Academic, New York, 1959, p. 317.

132. T. H. Gronwall, V. K. LaMer, and K. Sandved, *Phys. Z.,* **29**, 358 (1928).

133. T. H. Gronwall, V. K. LaMer, and L. Greiff, *J. Phys. Chem.,* **35**, 2245 (1931).

134. A. L. Loeb, P. H. Wiersema, and J. Th. G. Overbeek, *The Electrical Double Layer Around A Spherical Colloid Particle,* MIT Press, Cambridge, Mass., 1961.

135. P. H. Wiersema, A. L. Loeb, and J. Th. G. Overbeek, *J. Colloid Sci.,* **22**, 78 (1966).

136. J. Lijklema and J. Th. G. Overbeek, *J. Colloid Sci.,* **16**, 501 (1963).

137. D. Stigter, *J. Phys. Chem.,* **68**, 3600 (1964).

138. F. Booth, *J. Chem. Phys.,* **22**, 1956 (1954).

139. M. von Smoluchowski, *Kolloid Zh.,* **18**, 190 (1916).

140. F. Booth, *Proc. Roy. Soc., Ser. A* (London), **203**, 533 (1950).

CHAPTER 2

THE NUCLEIC ACIDS

Robert Steiner and David B. S. Millar

NAVAL MEDICAL RESEARCH INSTITUTE
BETHESDA, MARYLAND

I. GENERAL REMARKS

The importance of the nucleic acids to biology is well recognized (*1–5*). The problem of interpreting their physical properties in terms of their structure is a supremely interesting and challenging one.

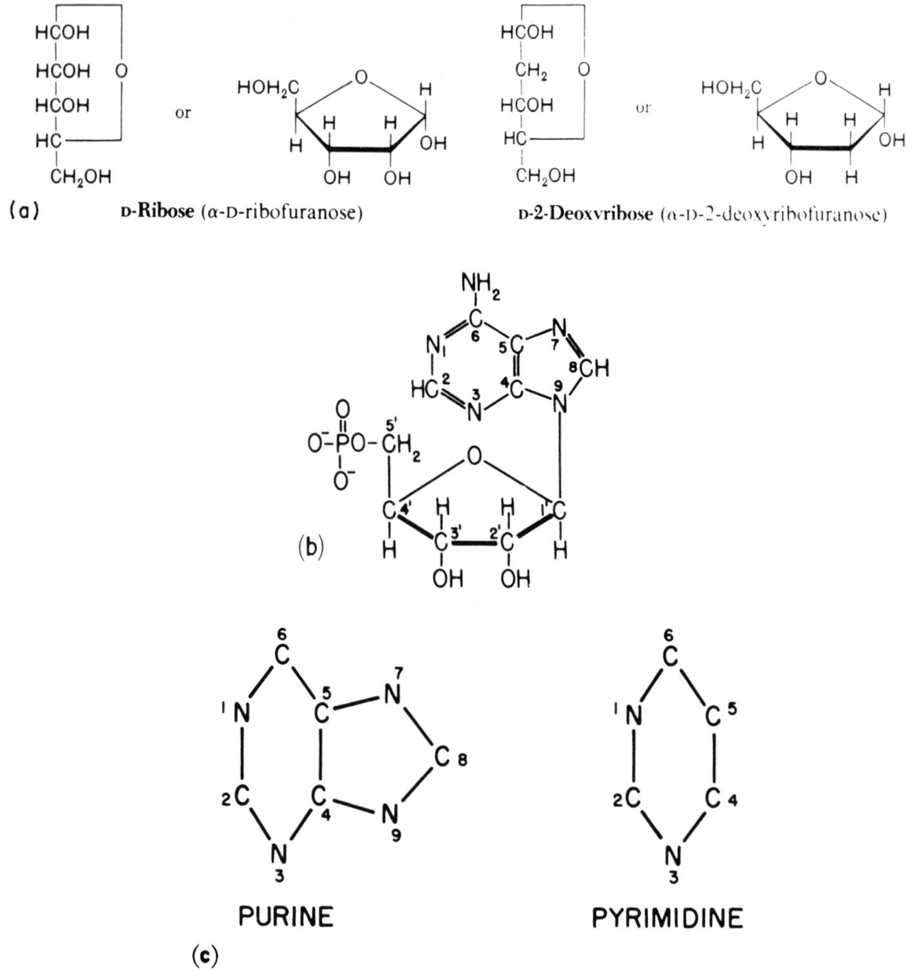

(a)   D-Ribose (α-D-ribofuranose)   D-2-Deoxyribose (α-D-2-deoxyribofuranose)

(b)

(c)   PURINE   PYRIMIDINE

Fig. 1. (a) Structures of D-ribose and 2-deoxy-D-ribose. (b) Structure of the ribonucleotide adenosine-5'-monophosphate (5'-AMP or 5'-adenylic acid). (c) Ring positions of the purine and pyrimidine bases, designated according to the *Fischer* system. The *International* system is equivalent for the purines, but for the pyrimidines N-3 becomes N-1 and the numbering proceeds clockwise around the ring, N-1 (Fischer) becoming N-3, C-6 becoming C-4, etc.

Both natural and biosynthetic nucleic acids fall within the scope of this chapter. The former include deoxyribonucleic acid (DNA), the universal carrier of genetic information for all cellular organisms and many viruses; ribosomal, messenger, and soluble ribonucleic acid (RNA), which occur in the cytoplasm of all living cells; and viral RNA. The latter include the DNA and RNA analogues produced by the enzymes *DNA-* and *RNA-polymerase* as well as the varied group of polyribonucleotides synthesized by *polyribonucleotide phosphorylase.*

The chemical aspects of nucleic acid structure have been adequately discussed elsewhere (*1,2*) and require only a brief summary here. All nucleic acids are linear, unbranched polymers of subunits called *nucleotides.* Each nucleotide consists of a purine or pyrimidine base; a 5-carbon sugar, which is D-ribose for RNA and 2-deoxy-D-ribose for DNA; and a

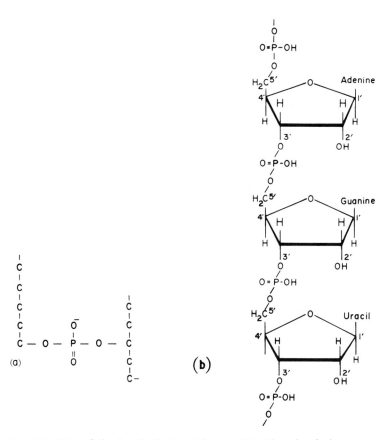

Fig. 2. (a) The 3′-5′ phosphodiester linkage. (b) The chemical structure of RNA. That of DNA is similar, except for the replacement of ribose by deoxyribose.

phosphate group (Fig. 1). The base is joined to the C-1' position of the sugar by a C—N β-glycosidic bond (Fig. 1), while the phosphate is esterified at the C-3' or C-5' position. All the natural and biosynthetic nucleic acids thus far examined consist of linear polymers of nucleotides united by phosphodiester linkages between the 3' and 5' carbons of adjacent nucleotides (Fig. 2).

The purine and pyrimidine bases of common occurrence in natural nucleic acids are five in number. The purines adenine and guanine and the pyrimidines uracil and cytosine occur in RNA (Fig. 3). In DNA uracil is replaced by thymine (Fig. 3).

A number of other bases occur in minor quantities or in special cases. In the DNA of the T-even bacteriophages cytosine is completely replaced by 5-hydroxymethylcytosine (6). 5-Methylcytosine is found in small quantities in the DNA of plants, especially wheat germ, as well as some higher animals (7). The base 6-methylaminopurine is of general occurrence in bacterial DNA's (8).

Bases other than the usual four are more numerous in RNA than in

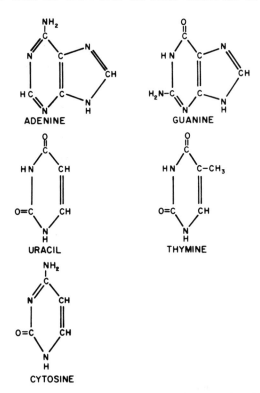

Fig. 3.   The bases of common occurrence in the natural nucleic acids.

DNA and are especially prominent in soluble RNA. These include, in addition to 6-methylaminopurine (9) and 5-methylcytosine (10), the bases 1-methyladenine (11), 2-methyladenine (7), 6,6-dimethylaminopurine (10), 1-methylguanine (12), 2-methylamino-6-hydroxypurine (12), and 2,2-dimethylamino-6-hydroxypurine (13).

TABLE I

A. Nomenclature of the Common Nucleotides and Their Polymers

Base	Position of phosphate	Name	Abbreviation	Polymer	Abbreviation
Ribonucleotide series:					
Adenine	3′	Adenosine-3′-monophosphate; 3′-adenylic acid	3′AMP	Polyriboadenylic acid	poly A; rA
	5′	Adenosine-5′-monophosphate; 5′-adenylic acid	5′-AMP	Polyriboadenylic acid	poly A; rA
Uracil[a]	5′	Uridine-5′-monophosphate; 5′-uridylic acid	5′-UMP	Polyribouridylic acid	poly U; rU
Cytosine[a]	5′	Cytidine-5′-monophosphate; 5′-cytidylic acid	5′-CMP	Polyribocytidylic acid	poly C; rC
Guanine[a]	5′	Guanosine-5′-monophosphate; 5′-guanylic acid	5′-GMP	Polyriboguanylic acid	poly G; rG
Deoxyribonucleotide series:					
Adenine[a]	5′	Deoxyadenosine-5′-monophosphate; 5′-deoxyadenylic acid	5′-dAMP	Polydeoxyadenylic acid	poly dA; dA
Thymine[a]	5′	Deoxythymidine-5′-monophosphate; 5′-deoxythymidylic acid	5′-dTMP	Polydeoxythymidylic acid	poly dT; dT
Cytosine[a]	5′	Deoxycytidine-5′-monophosphate; 5′-deoxycytidylic acid	5′-dCMP	Polydeoxycytidylic acid	poly dC; dC
Guanine[a]	5′	Deoxyguanosine-5′-monophosphate; 5′-deoxyguanylic acid	5′-dGMP	Polydeoxyguanylic acid	poly dG; dG

TABLE I (Continued)
B. Nomenclature of Nucleoside Diphosphates and Triphosphates

Base	Position of di- or triphosphate	Name	Abbreviation
Ribose series:			
Adenine[b]	5'	Adenosine-5'-diphosphate	ADP
Adenine[b]	5'	Adenosine-5'-triphosphate	ATP
Deoxyribose series:			
Adenine[b]	5'	Deoxyadenosine-5'-diphosphate	dADP
Adenine[b]	5'	Deoxyadenosine-5'-triphosphate	dATP

[a] Only the 5'-nucleotide is cited. The nomenclature of the 3'-nucleotide is entirely analogous.

[b] Only the adenine derivatives are cited here. The nomenclature of the nucleoside di- or tri-phosphates containing the other bases is entirely analogous.

In addition to the above, a number of derivatives of the common bases have been introduced into biosynthetic polynucleotides, either by direct incorporation or by chemical modification of the polymers. These include hypoxanthine, the deaminated derivative of adenine, and several halogenated derivatives of uracil.

The nomenclature of the common nucleotides and their polymers is summarized in Table I.

II. MOLECULAR ARCHITECTURE OF THE NUCLEIC ACIDS

Virtually all the natural and biosynthetic nucleic acids can be said to have some degree of organized structure. In the cases for which definitive information is available, the structure is invariably based upon a helix. The dramatic success of the doubly stranded helical model, initially proposed by Watson and Crick (*14*) and subsequently confirmed and refined by Wilkins and coworkers (*15*), in accounting for the properties of DNA has stimulated many investigators to search for analogous structural forms in RNA and the biosynthetic polynucleotides.

The Watson-Crick model will be considered in detail in a later section. In brief, the model consists of two anti-parallel polynucleotide chains wrapped as a double helix about a common axis. The purine and pyrim-

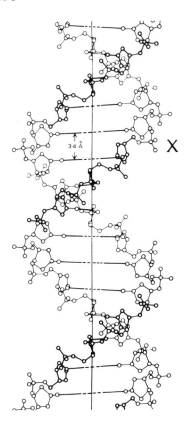

X

34 Å

Fig. 4. A schematic version of the molecular structure of DNA (B form), shown in projection. The planar bases have their edges toward the viewer, with the hydrogen bonding indicated by dashed lines.

idine bases are stacked in a parallel array in the core of the helix and are roughly perpendicular to the helical axis (Fig. 4). Each adenine is hydrogen bonded to a thymine in the other strand and each guanine to a cytosine (Fig. 5).

Since it was originally believed that hydrogen bonding contributed all or most of the stabilization energy for polynucleotide helices, considerable effort was directed toward establishing the sterically feasible hydrogen-bonded base pairs. It was found by Donohue that the Watson–Crick pairs by no means exhausted the list of possibilities (16).

Subject to certain assumptions as to the length and deviation from linearity of the hydrogen bonds, Donohue found a total of 24 hydrogen-bonded base pairs which are sterically feasible. These are cited in Table II. The numbers identifying the groups which act as donors or acceptors

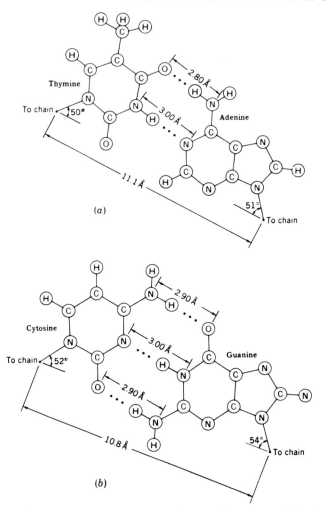

Fig. 5. Hydrogen-bonded base pairs occurring in natural DNA.

are shown in Fig. 6.   Uracil may be substituted for thymine in any of the pairs cited in Table II.

The hypoxanthine base of inosine is similar in structure to guanine, expect for the presence of an external amino group at the C-2 position in the latter base.   Hypoxanthine can replace guanine in a number of the pairings of Table II in which this group is not involved.   This is the case for pairs 7, 9, 13, 14, and 15.

The original Watson–Crick pairings correspond to 5 and 15 in Table II.   While the hydrogen bonding of the adenine–thymine pair is still regarded as that of 5, the guanine-cytosine pair is now believed to form an

TABLE II

Possible Hydrogen-Bonded Base Pairs

	Base I–Base II	Hydrogen bonding (I–II)
(1)	adenine–adenine	$d2$–$a3$; $a3$–$d2$
(2)	adenine–adenine	$d1$–$a1$; $a1$–$d1$
(3)	adenine–adenine	$d2$–$a1$; $a3$–$d1$
(4)	adenine–thymine	$d1$–$a2$; $a1$–d
(5)	adenine–thymine	$d1$–$a1$; $a1$–d
(6)	adenine–cytosine	$d1$–$a2$; $a1$–$d2$
(7)	adenine–guanine	$d1$–$a1$; $a1$–$d1$
(8)	adenine–guanine	$d1$–$a2$; $a1$–$d3$
(9)	guanine–guanine	$d1$–$a1$; $a1$–$d1$
(10)	guanine-guanine	$d1$–$a1$; $d2$–$a3$
(11)	guanine-guanine	$d1$–$a3$; $d2$–$a1$
(12)	guanine-guanine	$d3$–$a2$; $a2$–$d3$
(13)	guanine–thymine	$d1$–$a1$; $a1$–d
(14)	guanine–thymine	$d1$–$a2$; $a1$–d
(15)	guanine–cytosine	$d1$–$a2$; $a1$–$d2$
(16)	guanine–cytosine	$d1$–$a2$; $d2$–$a1$
(17)	guanine–cytosine	$d1$–$a1$; $d2$–$a2$
(18)	thymine–thymine	d–$a1$; $a1$–d
(19)	thymine–thymine	d–$a2$; $a1$–d
(20)	thymine–thymine	d–$a2$; $a2$–d
(21)	thymine–cytosine	d–$a2$; $a2$–$d2$
(22)	thymine–cytosine	d–$a2$; $a1$–$d2$
(23)	cytosine–cytosine	$d2$–$a2$; $a2$–$d2$
(24)	cytosine–cytosine	$d2$–$a1$; $a1$–$d2$

additional hydrogen bond (*17*) and to exist as the triply-bonded form $d1$–$a2$; $a1$–$d2$; $d2$–$a1$ (Fig. 5).

The list of feasible base pairs by Donohue was based solely on geometrical factors concerning the bases alone. Many of these pairings may be difficult to achieve because of steric difficulties associated with the ribose–phosphate backbone. Most have never been observed in actual polynucleotide systems.

Recent theoretical considerations developed by Zimm and others have led to considerable revision of the original concepts as to the origins of the stability of double and multistranded polynucleotide helices (*18*). It is believed today that hydrogen bonding makes only a secondary contribution to the thermodynamic stability of these systems. Basically, this is because an unbonded base can form hydrogen bonds with the surrounding water, so that there is no net gain in the number of hydrogen bonds when a base pair is formed. An improper base pair is unstable because it

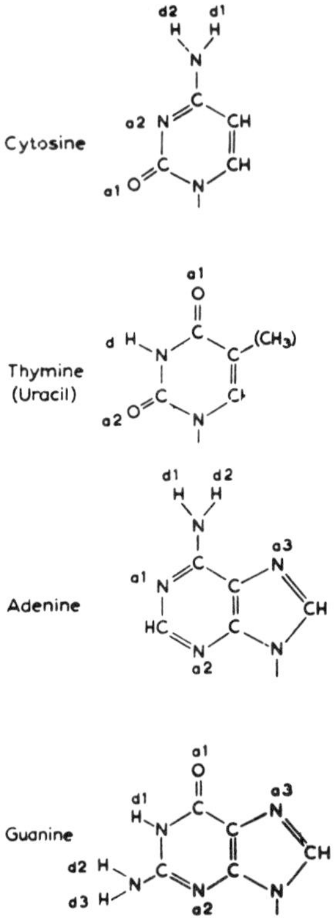

Fig. 6. Identification of the hydrogen bonding groups of Table II.

cannot form as many hydrogen bonds with water as the separated bases. Hydrogen bonding thus serves as a discriminator in permitting only the "correct" base pairs to form without incurring an energetic penalty in broken bonds to water.

Most of the free energy of stabilization is contributed by van der Waals interactions of the stacked bases. Crothers and Zimm have concluded that the base pairs of the DNA helix are stable with respect to separation by about -1 kcal of free energy, which includes contributions of -2 to -3 kcal from hydrogen bonding, -7 kcal of stacking free energy, and contributions from other, unresolved sources of $+8$ to $+9$ kcal (*18*).

Helical structures of the DNA type recur for other polynucleotide systems. However, while all the structures thus far observed are helical in nature, there are many systems whose helical organization is different from that of DNA. With regard to structure, the known natural and biosynthetic polynucleotides may be classified as follows:

(1) Systems devoid of organized structure. Only a handful of poly-nucleotides fall into this category, including polyribouridylic acid at ordinary temperatures ($>15°$) and polyriboinosinic acid at alkaline pH (>10).

(2) Single-stranded helices. Helical systems of this type are probably not hydrogen bonded and are stabilized entirely by van der Waals interactions. The alkaline forms of polycytidylic and polyadenylic acids appear to fall in this category.

(3) Doubly-stranded helices of the DNA-type. These are geometrically similar to the DNA helix and have similar hydrogen bonding patterns. Among the entries in this category, in addition to DNA itself, are the $dA:dT$ alternating copolymer produced by DNA-polymerase and the 1:1 complexes formed by polyadenylic acid with polyuridylic acid and by polycytidylic acid with polyinosinic or polyguanylic acid.

(4) Doubly-stranded helices of a non-DNA type. These include the acid forms of polyadenylic and polycytidylic acids.

(5) Multistranded helices formed by homopolynucleotides of two or more kinds. This category includes the triply-stranded species formed by polyadenylic acid with polyuridylic acid (A:2U) and with polyinosinic acid (A:2I), as well as the triply-stranded complex of polycytidylic acid with polyinosinic acid (C:2I).

(6) Multistranded helices formed by a single homopolynucleotide. The helical forms of polyinosinic acid and of polyguanylic acid are in this category.

(7) Irregular helical systems. This poorly defined category includes systems whose helical content is fractional, or which may contain more than one form of helical arrangement. Most types of RNA, as well as many biosynthetic nucleotide copolymers, may be placed in this class.

The presence of an organized molecular structure based on a helix is normally reflected by the occurrence of distinctive optical properties which are often used as a measure of helical content. These include:

(1) Hypochromism. The molar ultraviolet absorbancy per nucleotide of an organized polynucleotide is, in general, less than that predicted for an equivalent mixture of free nucleotides by a considerable factor, which depends upon wavelength. The various theories of hypochromism agree in attributing it to the mutual interactions of the electron systems of the bases, which are stacked in a parallel array.

(2) Optical rotation. The specific rotation of helical polynucleotides has a large positive component at wavelengths above 300 mμ, as compared with the individual nucleotides. At lower wavelengths a pronounced Cotton effect is observed. However, the optical rotatory dispersion has

not proved as useful for quantitative estimation of helical content as in the polypeptide case.   Thus the dispersion at visible wavelengths for many polynucleotides has been found to be described by a one-term Drude equation.

(3) Circular dichroism.   For helical polynucleotide systems the molar extinction coefficients are different for left- and right-handed circularly polarized light.   The difference, $\epsilon_L - \epsilon_R$, is termed the circular dichroism and provides a semiempirical index of helical content.   A structural transition from a helical to an unorganized form is reflected by a loss of circular dichroism.

III. STRUCTURELESS POLYNUCLEOTIDES

A. Polyribouridylic Acid

At the present time only one polynucleotide is believed to be largely devoid of organized structure at 25° and neutral pH.   Polyribouridylic acid (poly U or rU) has, at ordinary temperatures, the hydrodynamic and optical properties expected for a structureless, flexible polyelectrolyte. The principal evidence leading to this conclusion is as follows:

(1) The X-Ray diffraction patterns of fibers of poly U are of an amorphous character (19).   Moreover, electron microscopic examination has revealed only featureless coils.

(2) The hypochromism of poly U in the neighborhood of the primary absorption band at 260 mμ is very small—about 8 to 10%—and is almost unaffected by an increase in temperature or by urea (20).

(3) The optical rotation at wavelengths above 300 mμ does not contain the large positive component generally associated with helical content in polynucleotides (21–23).   Moreover, the optical rotatory dispersion is simple and can be fitted by a one-term Drude equation (24).

(4) The ionization of the uracil residues of poly U is essentially normal (24,25).   The pK is shifted only slightly to the alkaline as compared with the monomer (p$K \simeq 9.5$), and there is no sign of the cooperative titration behavior expected if the uracil groups were involved in extensive hydrogen bonding.

The preceding is virtually conclusive in ruling out any important degree of organized structure for poly U at ordinary temperatures.   The solution properties of poly U can thus provide a valuable "base line," giving an index of the behavior to be expected of a polynucleotide devoid of molecular organization.   It should however be stressed that these considerations do not apply to poly U at low temperatures ($<15°$), where a transition occurs to an organized helical structure (23). This will be discussed separately.

The intrinsic viscosity and the sedimentation coefficient of poly U preparations have been reported to show the following exponential dependence upon molecular weight (24) in 0.15 M NaCl (Fig. 7):

$$S_{20}^0 = 3.29 \times 10^{-2} M^{0.42}$$
$$[\eta] = 8.98 \times 10^{-5} M^{0.75}$$
(1)

Some question might be raised as to the quantitative validity of the above expressions, since the molecular weights used were average values for polydisperse preparations. Moreover, molecular weights were computed from the Flory–Mandelkern equation, which applies strictly only to a system of Gaussian coils (26) and does not yield a well-defined average molecular weight.

These limitations aside, the exponents of Eqs. 1 are typical of flexible randomly coiled chains. It can be concluded that the assumption of a randomly coiled configuration gives a self-consistent picture of the hydrodynamic properties of poly U.

The specific viscosity of a poly U preparation of molecular weight 10^5 has been found to be independent of shear rate (24). This behavior is likewise characteristic of randomly coiled polymers and precludes a highly asymmetric conformation.

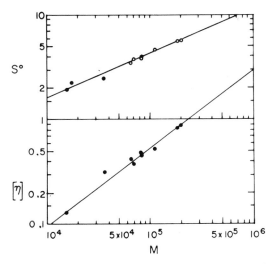

Fig. 7. Logarithmic dependence of sedimentation coefficient and intrinsic viscosity upon molecular weight for poly U in 0.15 M NaCl, 0.015 M Na citrate, pH 6.0 (24).

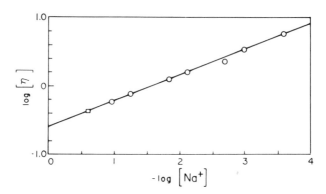

Fig. 8. Dependence of intrinsic viscosity of poly U upon NaCl concentration (24).

The specific viscosity of poly U at low polymer concentration decreases in a strictly linear manner with NaCl molarity (Fig. 8). The consistent linearity of the plot makes it unlikely that any abrupt conformational change occurs (24). The slope of the plot, 0.38, is typical of permeable polymer chains (27).

The molecular weight was found to be independent of ionic strength (24), suggesting that poly U remains single stranded at all ionic strengths at 25°.

An additional test may be applied for the internal consistency of the randomly coiled model. If this structural assumption is correct, the exponents of molecular weight in Eqs. 1 for sedimentation, a_s, and for viscosity, a_η, should obey the following relationship:

$$3a_s + a_\eta = 2 \tag{2}$$

This condition has been found to be satisfied (24).

Finally, the ionic strength dependence of viscosity can provide a clue as to the free-draining character of the polymer coil. (The velocity of solvent flow through a free-draining coil during a translational movement of the latter is equal to the relative velocity of the polymer with respect to the solvent. In other words, a free-draining coil does not immobilize solvent or impede its flow. The model is, of course, an abstraction which real systems may approximate more or less closely.) Cox has shown that the viscosity of a randomly coiled polyelectrolyte should vary as the 0.6 power of counterion concentration if it is impermeable to solvent, but as the 0.4 power if it is free-draining (27). Since the observed exponent is 0.38, there is a definite implication that poly U is free-draining, and hence has a relatively open structure with much internal space accessible to solvent

IV. SINGLE-STRANDED HELICAL POLYNUCLEOTIDES

A. Alkaline Polyriboadenylic Acid

The pH limits of stability of the alkaline form of polyriboadenylic acid (poly A or rA) are dependent upon the ionic strength and temperature. In 0.1 M KCl at 25° the transition to the acid form occurs at about pH 6 (28).

Alkaline poly A is definitely known to be single stranded (25,29). Nevertheless there is strong evidence for the presence of some form of molecular organization based on the helix. Among the more compelling data leading to this conclusion are the following:

(1) Poly A has a high degree of temperature-dependent hypochromism, amounting to about 40% at 260 mμ for a polymer of molecular weight 10^5 at 25° (20,30). The absorbancy increases gradually with increasing temperature (Fig. 9), with no sign of any abrupt transition (30). It was early recognized that the temperature dependence of hypochromism is very difficult to explain except in terms of the thermal disruption of some structure. The theories of Tinoco (31) and of Zimm and Kallenbach (32) have interpreted the hypochromism of helical polynucleotides as due to the interaction of the electron systems of neighboring bases stacked in a parallel array. To date no convincing molecular model which permits

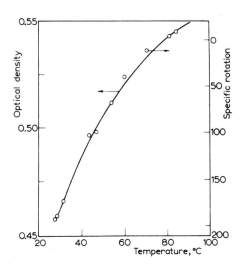

Fig. 9. Thermal profile of ultraviolet absorbancy at 257 mμ and of specific rotation at 589 mμ for poly A in 0.15 M citrate, 0.15 M NaCl, pH 7.0 (30).

such stacking has been proposed for a polynucleotide which is not based upon a helix.

(2) The optical rotatory and circular dichroism properties of alkaline poly A are suggestive of a helical organization (*30,33*). This kind of inferential argument, which is largely based upon analogy with known helical systems, does not of course constitute proof of such a structure, in view of the imperfect theoretical understanding of both phenomena.

The optical rotatory dispersion of alkaline poly A is anomalous (Fig. 10) and can be fitted with a two-term Drude equation (*33*).

Poly A at pH 7.4 (0.1 *M* NaCl) shows a circular dichroism curve composed of two parts (Fig. 11): a positive branch with a maximum at 264 mμ and a negative branch with a minimum at 247 mμ (*34*). This behavior is characteristic of helical polynucleotides (*34*). This form of poly A appears to be stable between 0 and 10°. The magnitude of the circular dichroism decreases progressively at higher temperatures, suggesting a gradual loss of ordered structure (*34*).

The optical rotation of poly A at wavelengths above 300 mμ contains a large positive component, in analogy to other helical polynucleotides and in contrast to poly U (*30*). The magnitude of the positive rotation decreases with temperature in synchrony with the absorbancy (*30*).

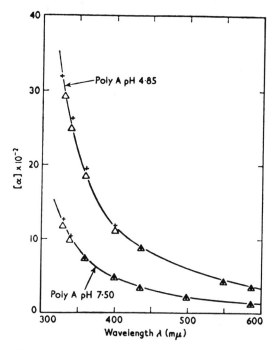

Fig. 10.   The optical rotatory dispersion of poly A at two pH's (*33*).

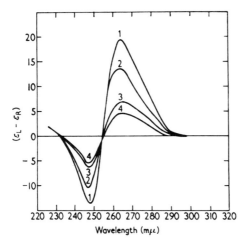

Fig. 11.   Circular dichroism curve of poly A in 0.1 M NaCl, 0.1 M tris, pH 7.4
(34).   Curve 1: 10°, curve 2: 33°, curve 3: 63°, curve 4: 73°.

(3) Low angle X-ray diffraction measurements of solutions of alkaline
poly A have indicated that regions of rodlike conformation are present,
with a radius of curvature of over 100 Å (35).   The mass per unit length
is half that of DNA, with one nucleotide per 3.4 Å.

The fragmentary information cited above does not permit more than the
assignment of a rather rough model for the molecular configuration of
alkaline poly A.   It seems clear that some form of helical organization,
which melts out gradually as the temperature is raised above 10°, is
present.   It was originally speculated that the helical regions might be
formed by the bending of the strand back upon itself to form hairpinlike
zones of organized structure, separated by structureless regions (30).
According to this model the helical regions would be doubly stranded and
somewhat analogous to DNA or the acid form of poly A.

However, the low angle X-ray diffraction results of Luzzati and coworkers (35) are definitely inconsistent with a model of this kind, inas-
much as the mass per unit length is only half that of DNA.   The best
tentative model for alkaline poly A (36) appears to be a single-stranded
helix stabilized entirely by van der Waals interactions of adjacent stacked
bases (Fig. 12).   Hydrogen bonding is probably not present to an important degree.   There is as yet no quantitative information as to the details
of the structure.

With regard to the overall molecular shape of alkaline poly A, the
prevailing evidence appears to favor strongly a randomly coiled configuration at 25°C.   This is not necessarily in conflict with the incorporation of
most of the molecule into regions of helical structure, provided that these

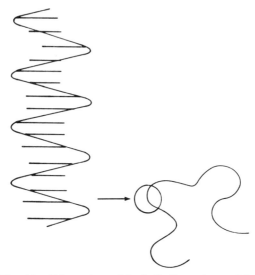

Fig. 12.   Schematic model of alkaline poly A (*36*).

are of limited extent and are separated by flexible, structureless zones or
that the helices are themselves appreciably flexible.

Light-scattering data have indicated that alkaline poly A is coiled to
$\frac{1}{13}$ or less of the maximum extension at ionic strengths of 0.01 or greater
(*37*).   This is considerably less than the ratio of $\frac{1}{3}$ which Treloar has
taken as a rough upper limit for the extension of Gaussian coils (*38*).

The dependence of sedimentation coefficient and intrinsic viscosity
upon molecular weight has been reported to be (*39*):

$$S = KM^{0.45}$$
$$[\eta] = KM^{0.65} \tag{3}$$

Both exponents are typical of random coils.

Poly A displays the usual expansion at low ionic strengths which is
characteristic of flexible polyelectrolytes.   Both the specific viscosity
and the radius of gyration, as measured by light scattering, increase with
decreasing concentration of electrolyte (*30,37*).

Further evidence for the flexibility of alkaline poly A is provided by
fluorescence polarization data (*40*).   The apparent rotational relaxation
time of poly A covalently labeled with the fluorescent dye acriflavine, as
determined by the procedures described by Weber and others (*41,42*), is
much too small to be consistent with the incorporation of the entire molecule into a single completely rigid rotational kinetic unit (*40*).

Eisenberg and Felsenfeld have recently succeeded in fractionating poly
A into preparations of a narrow molecular weight range and have examined

the properties of these fractions by viscosity and light scattering (*40b*). Both the intrinsic viscosity and the radius of gyration of alkaline poly A in 1 M NaCl, pH 7.5, were found to be very temperature-dependent.   The magnitudes of both parameters at 0°C indicated that, under these conditions, alkaline poly A has an ordered and extended configuration.   As the temperature is raised, the intrinsic viscosity and the radius of gyration decrease rapidly, approaching values characteristic of randomly coiled chains at 27°C.   At the latter temperature the intrinsic viscosity varies as the 0.5th power of molecular weight, as would be expected for a system of Gaussian coils.

Eisenberg and Felsenfeld have interpreted these results in terms of the following model.   At low temperatures alkaline poly A exists as an extended single-stranded helix, stabilized entirely by base stacking.   As the temperature increases a fraction of the bases lose, wholly or partially, their stacked orientation.   The molecule thereby acquires greater flexibility and collapses into a typical random coil, in which stacked and helical regions of limited extent persist.   The transition, as judged by the thermal profiles of absorbancy and viscosity, is essentially noncooperative in nature.

Perhaps the best over-all picture of alkaline poly A at ordinary temperatures is that of a molecule possessing extensive *short-range* order but *long-range* randomness of configuration.   The ordered regions consist of single-stranded helices, of uncertain dimensions, which are twisted about the helical axis.   The configuration is dynamic rather than static.

B. Alkaline Polyribocytidylic Acid

The alkaline form of polyribocytidylic acid (poly C or rC), which prevails for aqueous solutions at pH's above 7, resembles alkaline poly A in many of its properties.   A high degree of hypochromicity, which decreases gradually with increasing temperature (*43,44*), is present.   The thermal profile of absorbancy is essentially unchanged by pretreatment with formaldehyde (*44*).

The optical rotation of poly C has a large positive component at wavelengths above 300 mμ (*44*).   Moreover, an important positive Cotton effect occurs in the 270–300 mμ region (*44*), as shown in Fig. 13.   Alkaline poly C has also been found to show circular dichroism, whose intensity diminishes gradually with increasing temperature (*45*).

While a rigorous examination has not been carried out, the magnitudes of the radius of gyration and the intrinsic viscosity are comparable to those of randomly coiled polynucleotides of similar molecular weight (*37,46*), and there is little doubt that the gross configuration is of this type.   Like alkaline poly A, poly C exhibits a molecular expansion under

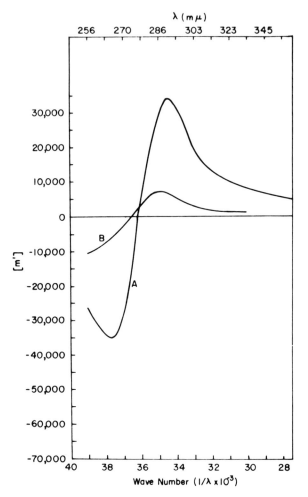

Fig. 13. Optical rotatory dispersion of poly C (curve A) and 5′-CMP (curve B) in 0.1 *M* citrate, pH 7.1 (*44*).

electrostatic stress, a major decrease in intrinsic viscosity occurring with increasing ionic strength (*46*).

The thermal profiles of optical rotation, ultraviolet absorbancy, and circular dichroism are gradual and do not show the sharpening character-istic of helical polynucleotides of the DNA type (*43–45*). The thermally-induced helix → coil transition is thus qualitatively similar in form to that of alkaline poly A.

Exhaustive treatment of poly C with formaldehyde, which blocks the primary amino groups, does not qualitatively change either the form of the thermal denaturation curve or the character of the optical rotatory dispersion (*44*). Fasman and coworkers have concluded that the helical structure is not stabilized by hydrogen bonding involving the amino groups, but rather by van der Waals interactions of stacked bases (*44*).

There is no quantitative information as to the dimensions of the helical regions. And as in the case of alkaline poly A, it is uncertain whether the flexibility of the strands is due to frequent interruptions of the helices by structureless segments or to an intrinsic flexibility of the poly C helix itself.

V. DOUBLY-STRANDED HELICAL POLYNUCLEOTIDES—DNA

A. Molecular Structure

The classical helical duplex model for DNA, which was postulated by Watson and Crick (*14*) and subsequently refined by Wilkins and coworkers (*15,47,48*), provides a logical point of departure for a discussion of the remaining polynucleotides of organized structure. The X-ray diffraction studies of Wilkins and his collaborators have established that oriented fibers of DNA can exist in several crystalline modifications which give appreciably different diffraction patterns and which differ structurally to a minor degree. All are variants of the basic Watson–Crick structure (Section V).

In all forms of DNA the two strands are, apart from base sequence, equivalent in configuration, but opposite in direction, or *anti-parallel* (*49*). That is, the sequence of atoms in the sugar–phosphate backbone (—C4'—C5'—O—P—O—C3'—) of each strand is the reverse of that of its partner. Rotation of the molecule by 180° does not alter its appearance.

At humidities less than about 70% the sodium salt of DNA crystallizes in the A form (*47*). This is very highly regular and crystalline, existing as a face-centered monoclinic lattice. The dimensions of the unit cell are: $a = 22.2$ Å, $b = 40.0$ Å, $c = 28.1$ Å. The unit cell contains a repeat unit of two DNA molecules, with the helical axis in the c direction.

In this form the bases are not strictly perpendicular to the helical axis, being tilted about 20°. The lattice contains about 40% water.

At higher humidities a transition occurs to the less crystalline B form (*48*). This appears to be the common form of DNA in biological systems and in aqueous solution. In this modification the bases are normal to the helical axis and are closer to the axis than in the case of the A form. The sugar ring is placed so that its plane is inclined as much as possible to the helical axis (Fig. 14). The base pairs, of which there are 10 for each turn of the helix, are spaced 3.4 Å apart. The sugar–phosphate chain is stretched almost to its limit and the phosphate groups are about 7 Å apart along the helix. The phosphorus atom is about 9.6 Å from the axis and the average diameter of the molecule is close to 20 Å. The two polynucleotide strands are separated on the surface of the molecule by two helical grooves.

The lithium salt of DNA has been found to crystallize in the B modification at low humidities, instead of undergoing the transition to the A form (47,49). This has permitted the determination of the dimensions of the unit cell, which are, at 66% relative humidity: $a = 22.72$ Å, $b = 31.28$ Å, $c = 33.60$ Å.

A third variant of the basic DNA structure, the C form, arises for the lithium salt upon reducing the relative humidity from 66 to 44% (50). This resembles the B form, except that the base pairs are moved about 2 Å away from the helical axis and tilted by 5°.

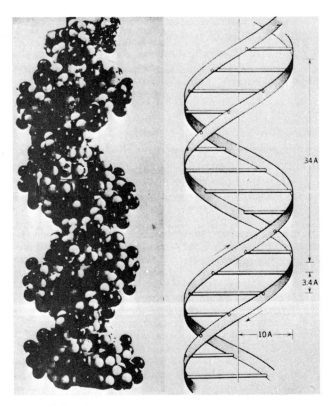

Fig. 14. Model of the B form of DNA.

Since the prevailing evidence indicates that the B form of DNA predominates in solution, the discussion to follow may be tacitly assumed to apply to this structure.

B. Size and Shape Parameters of DNA

Most of the available hydrodynamic information upon doubly-stranded helical systems has been obtained for DNA itself. It is now recognized that DNA of high molecular weight is very susceptible to degradation by the mechanical shearing stress encountered during isolation by routine procedures. For this reason almost all of the molecular weights which have been reported for DNA from mammalian or bacterial sources are probably much too low and should be regarded as essentially lower limits. The usual range of reported values is $5-15 \times 10^6$ (51). In the case of the DNA from T2 bacteriophage, after isolation by very mild procedures which avoided the difficulties cited above, a molecular weight close to 1.3×10^8 has been found.

While most of the physical measurements upon DNA solutions refer to degraded preparations, such studies have been useful in establishing the general shape characteristics of the molecule. Doty, McGill, and Rice have examined the sedimentation, viscosity, and light-scattering behavior of thymus DNA which had been subjected to ultrasonic degradation (52). There is strong evidence that ultrasonic treatment of DNA results in simultaneous cleavage of both strands, with little concomitant denaturation or single-chain scission. Sonic degradation thus provides a means of preparing intact DNA fragments of varying average length.

By using this approach these workers found that the sedimentation coefficients and intrinsic viscosities of the fragments showed the following exponential dependence upon weight-average molecular weight.

$$S_{20}^0 = 0.063 M_w^{0.37}$$
$$[\eta] = 1.45 \times 10^{-6} M_w^{1.16}$$

$$(4)$$

The exponent of M_w in the expression for $[\eta]$ is 1.16, which exceeds slightly the rough limiting value (1.0) for randomly coiled, solvent immobilizing polymers. The exponent of M_w in the equation for S_{20}^0, 0.37, is appreciably larger than the rough lower limit for random coils (0.33).

Both relations are clearly inconsistent with a model which represents DNA in solution as completely extended rigid rods. Such a model would predict a smaller exponent for S_{20}^0 and a larger one for $[\eta]$.

Parallel light-scattering measurements established the relationship between the radius of gyration and the molecular weight (52). This was found to be of the form:

$$R_G = 8.3 \times 10^{-3} M_w^{0.5} \qquad\qquad (5)$$

Again the exponential dependence is quite different from expectations for a system of rigid rods, and in this case it agrees with that predicted for random coils.

Light-scattering measurements upon undegraded DNA are subject to some uncertainty of interpretation because of the difficulty of making a reliable extrapolation of reduced intensity to zero angle (53). If this is disregarded, the existing data indicate that the radius of gyration is much smaller, by a factor of four or more, than that expected for a *rigid* helix of the given molecular weight whose end-to-end separation is equal to the contour length.

Qualitatively, the model which best brings all these observations into harmony is that of a *stiff* coil with a large persistence length, whose behavior is intermediate to the rod and coil extremes (54). This model has been formalized as the *wormlike* chain of Porod (55).

The detailed angular dependence of light scattering by DNA solutions has been interpreted quantitatively in terms of this model (54). The computed persistence lengths (Table IIIA) are quite large and indicate a high degree of stiffness of the coils.

TABLE IIIA

Molecular Weights, Contour Lengths (L), and Persistence Lengths (a) of Two Samples of Thymus DNA as Determined by Light Scattering (54)

M_w	R_G, Å	L/a	L, Å	a, Å
6.6×10^6	2200	300	66,000	220
5.5×10^6	2300	90	38,700	430

The rigidity of DNA is sufficient to prevent the usual expansion under electrostatic stress characteristic of flexible polyelectrolytes. Thus the intrinsic viscosity of DNA is essentially independent of ionic strength within wide limits (56).

C. The Denatured State of DNA

The doubly-stranded helical structure of native DNA is stable only within definite limits of pH, ionic strength, and temperature. Outside these limits a transition occurs to *denatured* DNA, in which the helical organization is partially or completely lost. While the process appears to be intrinsically reversible, the reversal encounters severe kinetic barriers under many conditions. This often results in the persistence of the denatured state after a return to an environment where the native state is thermodynamically stable (57).

Denaturation was early shown to occur at extremes of pH at room temperature. The protonation at acid pH of adenine, cytosine, and guanine or the dissociation at alkaline pH of a hydrogen ion by thymine and guanine results in the rupture of one or more of the hydrogen bonds which stabilize each of the base pairs of the double helix. A second, and perhaps more important, destabilizing factor is the intense electrostatic repulsion between similarly charged bases packed into the core of the helix (58).

The earliest evidence for the acid and alkaline denaturation of DNA was derived from hydrogen ion titration data. If a solution of native DNA is titrated from neutral pH to either pH 2 or pH 12 and then back to neutrality, the forward and reverse titration curves do not coincide (59). The change for the reverse half cycles is in the direction of increased hydrogen ion binding at intermediate pH's for the acid branch and to increased dissociation for the alkaline branch (Fig. 15). Formally, the change corresponds to a shift of the base pK's toward neutrality after exposure to extremes of pH.

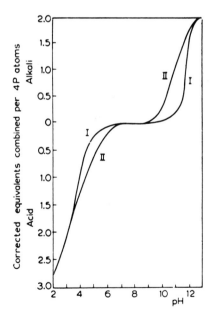

Fig. 15. Hydrogen ion titration curves of DNA at 25°. Curve I is the native DNA which is brought to acid or alkaline pH for the first time. Curve II is that obtained for DNA upon back titration from extremes of pH (59).

It has recently been reported (61b) that spectrophotometric titrations at different wavelengths indicate that over half of the cytosine bases of DNA can be protonated in 0.02 M KCl at 25° without extensive destruction of the helical structure. However, changes in optical rotatory dispersion suggest that some change in helical conformation occurs.

The degree of hysteresis depends strongly upon conditions. In general, the zone of pH in which titration is quantitatively reversible is extended to more acid values with increasing ionic strength and decreasing temperature (60,61). Cox and Peacocke, working with herring sperm ƆNA, have found that the divergence between forward and reverse branches is much less if an acid titration is made at 0.4° than if it is done at 25° (60,61). If titration is performed at −0.4°, the forward and reverse curves are identical down to pH 2.7 (60,61).

The appearance of irreversibilty in the titration curve is accompanied by profound changes in the hydrodynamic properties. The intrinsic viscosity drops by a factor of 10 or more (62). Parallel alterations occur in the sedimentation coefficient (63). Studier, working with coliphage DNA, has found that the sedimentation coefficient in 0.1 M NaCl is constant between pH 3 and pH 11.5 (63). Outside these limits a rapid increase occurs, followed by an abrupt decline at still lower, or higher, pH's (63). The maximum value of the sedimentation coefficient attained at acid pH was about twice that of the native molecule at neutral pH. The maximum was at about pH 2.7. The alkaline maximum occurred at pH 11.7, the value of sedimentation coefficient increasing by 50% over the value at neutral pH. This behavior has been interpreted as reflecting an initial collapse of the helical structure into a more or less randomly coiled form, followed by separation of the strands.

A return to neutral pH from pH 13 or pH 2.5 results in a sedimentation coefficient greater by a factor of two than that of native DNA under the same conditions. In contrast to native DNA, the sedimentation coefficient is now pronouncedly ionic strength-dependent, increasing with increasing ionic strength (Fig. 16). Denatured DNA appears to show the contraction with increasing electrolyte concentration characteristic of flexible polyelectrolytes. However, a parallel increase in hypochromism with increasing ionic strength indicates that base–base interactions are also an important factor.

The dependence of sedimentation coefficient upon molecular weight is altogether different from that of native DNA (63). For alkaline-denatured DNA

$$S^0_{20,w} = 0.0528M^{0.400} \qquad (6)$$

and after a return to neutral pH

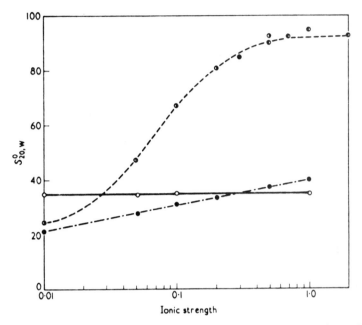

Fig. 16. Sedimentation coefficient of coliphage DNA as a function of ionic strength (63): (○) native DNA; (●) alkaline DNA; (◑, ◐) netural denatured DNA.

$$S_{20,w}^{0} = 0.0105 M^{0.549} \tag{7}$$

Both types of exponential dependence are typical of flexible random coils. This behavior is not of course inconsistent with the presence of numerous regions of short-range order.

Denaturation of DNA also occurs at elevated temperatures (64). The process has usually been monitored by the increase in absorbancy at 260 mμ reflecting the loss of hypochromism which accompanies the transition to the denatured form (64). By this criterion the transition is quite sharp, being generally 90% complete over a temperature interval of 5°. The midpoint of the thermal transition depends upon the base composition, increasing with the guanine plus cytosine content (Fig. 17).

Exposure of DNA to temperatures above the transition zone, followed by *rapid* cooling to 25°, results in a product whose solution properties resemble those of acid- or alkaline-denatured DNA and contrast vividly with those of native DNA (65). The intrinsic viscosity falls by a factor of 10 or more, while the radius of gyration, as determined by light scattering, falls several fold (54). Moreover, both the viscosity and the radius of gyration now become ionic strength-dependent (51).

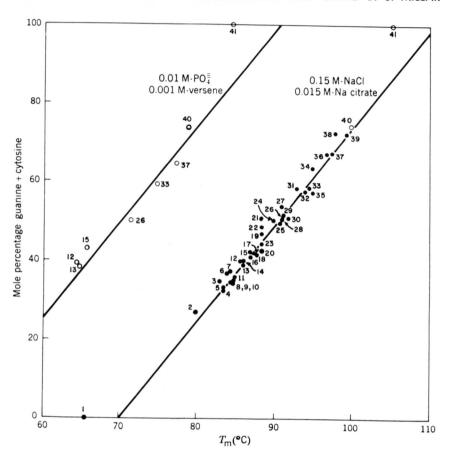

Fig. 17. Dependence upon base composition of the melting point of DNA's from various sources.

The issue of whether an actual separation of strands accompanies denaturation is unfortunately somewhat confused in the case of natural DNA, although the bulk of the current evidence appears to favor this interpretation. In the case of *E. coli* DNA in which one strand is labeled with N^{15}, clearcut evidence for a strand separation has been obtained by CaCl density gradient ultracentrifugation (*66*). Application of the Flory–Mandelkern equation for the computation of molecular weights from the sedimentation coefficient and the intrinsic viscosity has also indicated that the molecular weight of Pneumococcus DNA halves upon thermal denaturation (*67*). In contrast, most light-scattering measurements have failed to show a drop in molecular weight (*68*). The use of light scattering to determine molecular weights of asymmetric polymers of this size encounters practical difficulties because of the lengthy extrapolations of data required (53).

Harpst, Krasna, and Zimm (68b) recently reported that accurate values of molecular weight can be obtained for DNA by light scattering only if measurements are extended to very low angles with respect to the incident beam (~10°). If this is done, molecular weights obtained by light scattering for native DNA agree with those computed from combined sedimentation and viscosity data. Since the earlier light-scattering values were obtained from data measured at high angles (>30°), it is possible that the failure to observe a halving of molecular weight upon denaturation arises from inaccuracies originating from this factor.

Since fairly conclusive evidence for strand separation exists in the case of the helical complexes of several biosynthetic polynucleotides, it is probably justifiable to assume, as a working hypothesis, that it occurs also for native DNA, pending the availability of more definitive information.

Thermally denatured DNA which has been rapidly cooled to room temperature retains some degree of hypochromism, which is lost upon heating (64). However, the thermal profile no longer shows the sharp transition characteristic of native DNA (64). Instead a gradual increase of absorbancy with temperature is observed (Fig. 18). This kind of behavior is consistent with the partial reformation of base pairing, so that some degree of short-range order is present (64). A collection of ordered regions of limited extent would be expected to undergo the gradual thermal transition characteristic of denatured DNA (64).

Extensive reformation of the native structure of bacterial DNA occurs

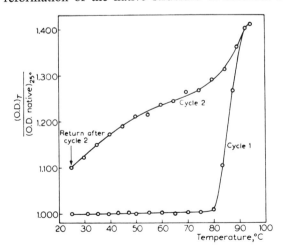

Fig. 18. The thermal dependence of absorbancy at 259 mμ for DNA taken through two heating cycles (64).

provided that cooling to room temperature is carried out gradually with prolonged standing at temperatures immediately below the transition zone (67). Under these conditions the formation and rupture of base pairings is rapid enough to permit the selective formation of the most stable structure, which is the native doubly-stranded helix. The process is often termed "annealing." DNA renatured in this way recovers its original properties, including the sharpness of its thermal transition and, in the case of transforming principle, its biological activity.

Several detailed studies of the renaturation kinetics of DNA have recently appeared (67b,67c). If the degree of hypochromism at 260 mμ is used as a measure of the extent of base pair reformation, the reaction has been found to obey second-order kinetics, at least up to degrees of renaturation of about 75% (67c). The thermal profile of the apparent rate constant is bell-shaped, showing a flat maximum at temperatures about 25° below the melting point. At constant temperature the rate decreases with increasing solvent viscosity.

Wetmur and Davidson (67c) have interpreted the second-order kinetics of the renaturation process as indicating that the rate-limiting step is the bimolecular interaction of two small complementary sequences in two different strands to form a nucleus. The nucleation step is followed by a rapid zippering up of the balance of the molecule to form the complete double helix. This kind of model has been shown to be consistent with the observed temperature dependence of the rate.

VI. DOUBLY-STRANDED HELICAL POLYNUCLEOTIDES OF THE DNA TYPE

A. The 1:1 Complex of Polyriboadenylic Acid and Polyribouridylic Acid

The first example to be observed of an artificial two-stranded helical polynucleotide was the 1:1 complex formed by polyriboadenylic and polyribouridylic acids. The 1:1 (rA:rU) complex forms spontaneously upon mixing solutions of its components at neutral pH and an ionic strength of 0.01 or greater (69,70). Under these conditions its formation is stoichiometric at U:A mole ratios of 1.0 or less, while at U:A ratios greater than 1.0 a transition occurs to the three-stranded rA:2rU species, formation of which is complete at a U:A ratio of 2:1 (70–72).

The reaction is most readily followed by the increase in hypochromism which accompanies it. The decrease in absorbancy at 260 mμ appears to be the same for the rA:rU and rA:2rU species, so that a plot of the absorbancy at 260 mμ as a function of the mole fraction of poly U has the form of two straight lines intersecting at a mole fraction of poly U equal to 0.67 (Fig. 19). However, the change in absorbancy at 280 mμ is very different for the two forms, and mixing curves obtained at this wavelength permit the unequivocal differentiation of the two stepwise reactions (Fig. 19) and indicate that, at 25°, the rA:rU species is formed at low U:A ratios and adds a second poly U strand at high U:A ratios (72). While this version of the mechanism has been challenged (73), it has been confirmed by infrared measurements (74) and is at present generally accepted.

The combination of poly A and poly U results in pronounced changes in the solution properties. The viscosity, sedimentation coefficient, and molecular weight all show major increases (69–71). The rotational relaxation time, as measured by fluorescence polarization for an acriflavine conjugate of poly A or poly U, also increases (40).

Fibers of what was assumed to be the rA:rU species have been examined by X-ray diffraction (75,76). While a detailed analysis was not made, the general similarity of the pattern to that of DNA led to the structural assignment of a two-stranded antiparallel helix of the DNA type, in which the hydrogen bonding between adenine and uracil was equivalent to that postulated to exist between adenine and thymine in DNA.

While this structure continues to be generally accepted, some doubts have recently been expressed in view of the finding that several crystalline

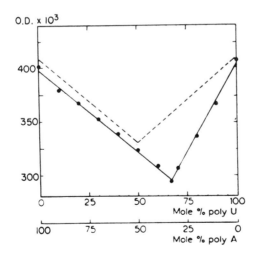

Fig. 19. The equilibrium absorbancies at 259 mμ for mixtures of poly A and poly U in 0.1 M NaCl, 0.01 M glycyl–glycine, and 0.0012 M MgCl$_2$, pH 7.4 (70). The dashed line shows the shape of the mixing curve for the formation of the rA:rU species only.

complexes of adenine and uracil derivatives have a different mode of hydrogen bonding, in which the adenine amino nitrogen is bonded to the C-2 carbonyl of uracil and the uracil N-1 to the adenine imidazole nitrogen N-7 (77). Although this alternative bonding scheme has been excluded for DNA itself and for the dA:dT copolymer, it has not been eliminated in the case of the rA:rU complex. A reinvestigation of this question would clearly be desirable.

The detailed kinetics of the combination, as monitored by the decrease in ultraviolet absorbancy, have been studied by Ross and Sturtevant (78) using a stopped-flow absorption spectrophotometer. At an ionic strength of 0.1 and 25° the change in absorbancy is complete within a few seconds. The kinetics are complex, being initially of second-order form and subsequently deviating toward first order. This kind of behavior is suggestive of an initial nucleation reaction in which multiple helical loci are formed, followed by the expansion of the helical nuclei. The rate decreases with temperature as the melting point is approached, becoming zero at the melting point (78). The second-order rate constant is strongly dependent upon ionic strength in the direction anticipated for a reaction between ions of similar charge.

When poly A and poly U are mixed at neutral pH, 25°, and an ionic strength of 0.1 or greater to form the rA:rU species, both the molecular weight, as measured by light scattering, and the ultraviolet absorbancy attain their final values very quickly (71). However, the intrinsic viscosity and the radius of gyration show a slow subsequent increase, suggesting that some degree of annealing of the complex species occurs, which results in the elimination of residual imperfections.

The rA:rU complex undergoes a thermal denaturation essentially similar to that of DNA (Fig. 20). At concentrations of NaCl less than about 0.2, the transition is from the two-stranded species directly to free poly A and poly U (72). At higher levels of NaCl, a disproportionation occurs at temperatures below the melting point:

$$2(rA:rU) \rightarrow rA:2rU + rA \tag{8}$$

This is followed at higher temperatures by a melting of the triply stranded form to the free components (72).

The rate of combination of poly A and poly U falls off rapidly at pH's below that at which poly A undergoes a transition to its acid form, becoming zero at pH 5.3 (71). If the preformed complex is titrated to acid pH, dissociation, as measured by an increase in absorbancy at 259 mμ, does not become evident until below pH 4.5 (71). The rA:rU complex also dissociates at alkaline pH upon ionization of the uracil groups (71). The transition is remarkably sharp, as measured by ultraviolet absorbancy or

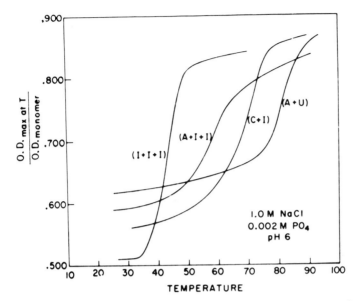

Fig. 20. Thermal transition for several helical polyribonucleotide complexes, as monitored by ultraviolet absorbancy (*64*).

fluorescence polarization, attaining completion over less than 0.1 pH unit (*40,71*).

The enthalpy change resulting from the formation of the rA:rU species has been measured calorimetrically (*79,80*). In 0.1 M KCl at 25° it is close to —6 kcal/base pair. This value includes a term of opposite sign arising from the loss of the helical structure of alkaline poly A. This has not been measured directly, but has been estimated indirectly as about 1 kcal for the above conditions. The corrected enthalpy is —7 kcal for the formation of an adenine–uracil base pair at 25°.

B. The Alternating dA:dT Copolymer

The DNA-polymerase from *E. coli* will, in the absence of added primer DNA, catalyze the synthesis of certain polydeoxyribonucleotides after an initial lag period. One of these, the dAT copolymer, which is formed if only dATP and dTTP are present, is of particular interest and has been intensively studied. The only bases present are adenine and thymine, which alternate with perfect regularity along the polynucleotide chains (*81*), so that the base sequence of each strand is ···ATATATAT···.

X-ray diffraction studies of the lithium salt of the dAT copolymer have shown that its structure is a double helix resembling the B form of natural DNA (*82*). For many purposes the dAT copolymer may be regarded as a biosynthetic analogue of DNA with a known and regular structure, in which the only base pairing is adenine–thymine.

The dAT copolymer undergoes a helix → coil transition upon heating. As in the cases of the other helical polynucleotides, the process may be monitored by observations of the ultraviolet absorbancy, a sharp decrease in hypochromism at 260 mμ occurring in the transition region (83). By this criterion the process has many points of resemblance to the analogous transition for DNA, but differs from it in several respects, including its complete and rapid reversibility (83). This relative ease of reversal is presumably a consequence of the alternating nature of the copolymer, which eliminates kinetic barriers to recovery arising from base mismatching.

A comparison of the thermal profiles of absorbancy and of viscosity reveals interesting differences (83,84). A drop in viscosity, which is especially pronounced at high ionic strength, begins at temperatures well below the onset of a perceptible change in the absorbancy (84). In 0.5 M NaCl a decrease of almost tenfold was observed prior to the optical transition. The drop in viscosity is followed by a rise in the optical melting zone (Fig. 21).

The limiting viscosity approached at temperatures well above the transition exceeds that of the intact dAT:dAT helix at low ionic strength (0.01 M NaCl) but is less than that of the helical form at high ionic strength (0.5 M NaCl). This effect arises from a pronounced dependence upon ionic strength of the intrinsic viscosity of the individual dAT strands, which reflects their flexibility and capacity to expand under electrostatic stress. In contrast, the intrinsic viscosity of the relatively rigid dAT:dAT

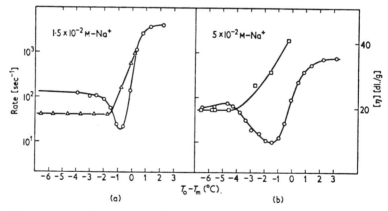

Fig. 21. A comparison between the melting rates [△, (a); □, (b)], as measured from the ultraviolet absorbancy using a thermal jump technique, and the intrinsic viscosity (○) of the dAT copolymer at temperatures (T_o) in the vicinity of the midpoint (T_m) of the transition (84).

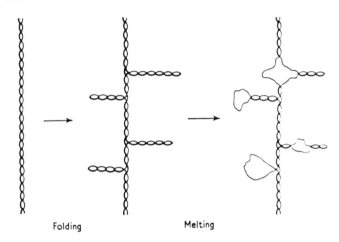

Folding   Melting

Fig. 22.   The model of Spatz and Baldwin (*84*) for the folding of the dAT copolymer in the pre-melting zone.

helix is essentially independent of electrolyte concentration within wide limits (*83,84*), in analogy to the behavior of natural DNA.

Several possible explanations can be advanced for the drop in viscosity at temperatures below the optical transition.   The most obvious of these, that internal interruptions in the helix produce flexible "hinge points," has been rejected on kinetic grounds (*84*).

Spatz and Baldwin (*84*) have favored an alternative mechanism which postulates a folding of the two-strands into hairpinlike exterior branches in the pre-melting zone (Fig. 22).   The alternating character of the dAT copolymer makes this plausible.   The branches subsequently melt independently.

Spatz and Baldwin have examined the kinetics of melting by a temperature jump method (*84*).   For a fixed final temperature within the melting zone the rate of melting was constant and independent of initial temperature for initial temperatures below the region of folding.   A rapid increase in rate was observed for initial temperatures within the folding zone, which was attributed to the initiation of melting at the exterior branches.

C. The dAT:dA$\overline{BU}$ Hybrid

Inman and Baldwin (*85*) have described the formation and properties of a hybrid helical complex, one of whose strands contains adenine and thymine in alternating sequence and the other adenine and bromouracil ($\overline{BU}$).   The dA$\overline{BU}$ copolymer may be synthesized enzymically in the same manner as dAT:dAT, with the replacement of dTTP with dB$\overline{U}$TP in the reaction mixture.   It appears to be generally similar in structure, existing as a two-stranded helix in which the base pairing is between adenine and bromouracil.

Formation of a true dAT:dA$\overline{BU}$ hybrid requires the heating of a mixture of dAT:dAT and dA$\overline{BU}$:dA$\overline{BU}$ through the thermal transition under conditions of high salt (0.5 M Na citrate), where the melting points of the two species are almost identical, followed by *slow* cooling. The extensive formation of the hybrid species requires a relatively high concentration (A_{260} = 30 to 100).

Since the densities of the dAT and dA$\overline{BU}$ strands are significantly different, Cs_2SO_4 gradient centrifugation could be used to follow the separation of the strands of the hybrid at temperatures above the transition zone.

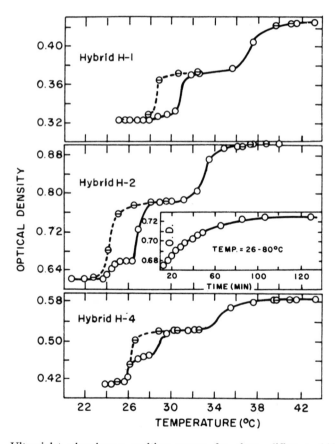

Fig. 23. Ultraviolet absorbancy melting curves for three different dAT:dA$\overline{BU}$ hybrid preparations in 0.002 M Na citrate (*85*). The inset shows the time dependence of absorbancy at a fixed temperature for H-2: ($\bigcirc$) original heating, ($\ominus$) second heating.

In a solvent of ionic strength 0.006 at 44°, which is slightly above the melting point for dA$\overline{BU}$:dA$\overline{BU}$, strand separation could be readily observed after rapid cooling to 25°.   The kinetics of separation were peculiar in that 60–80% of the hybrid was dissociated within 5 min at 44°, while the balance required up to 17 hr for complete dissociation.   No obvious explanation has been forthcoming.   It is possible that the dependence of the rate of unwinding upon strand length is at least partially responsible, in view of the probably wide distribution of lengths.

The thermal denaturation of dAT:dA$\overline{BU}$ is complicated by the fact that the products of its dissociation can themselves associate to form helical duplexes.   In low salt (0.002 M Na phosphate) the melting point of the hybrid is 3° above that of dAT:dAT and 6° below that of dA$\overline{BU}$: dA$\overline{BU}$.   At the low concentrations employed for ultraviolet absorbancy measurements ($A_{260} < 1$) the melting of the hybrid is not reversible.   The thermal profile of absorbancy is of a stepwise character (Fig. 23).   The initial step, which reflects the melting of the hybrid, is followed by the melting of the dA$\overline{BU}$:dA$\overline{BU}$ species which is subsequently formed.   Because of the irreversibility of the hybrid melting, a pronounced hysteresis is observed upon cooling from the hybrid transition zone.

D. The 1:1 Complex of Polycytidylic and Polyinosinic Acid

The first example to be reported of a complex of this type was the 1:1 complex between polyribocytidylic acid and polyriboinosinic acid (rC:rI) studied by Davies and Rich (86).   The ultraviolet absorption spectra of mixed solutions of the two homopolymers at neutral pH and ionic strength 0.1 were found to differ markedly from those predicted from simply additivity.   The spectral alteration corresponded to an increase in the degree of hypochromism at most wavelengths (Fig. 24).

If the absorbancy at 235 mμ was taken as a direct measure of the extent of reaction, the mixing curve for the above conditions consisted of two straight lines intersecting sharply at a 1:1 mole ratio (86).   An argument analogous to that applied earlier to the rA:rU case indicated that the reaction is reversible and stoichiometric and that the complex is two-stranded (or conceivably four-stranded).   There was no evidence for the formation of an rC:2rI or 2rC:rI species.

The formation of the rI:rC species could also be detected in the ultracentrifuge.   At neutral pH and ionic strength 0.1 the sedimentation coefficient of the 1:1 complex ($S^0_{20} = 13.9$) was much greater than those of the separate rI ($S^0_{20} = 7.4$) or rC ($S^0_{20} = 5.5$) species (86).

As in the case of the rA:rU system, the rate of formation of the complex species was governed by the ionic strength.   For the above prepara-

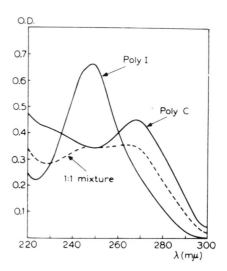

Fig. 24. The ultraviolet absorption spectra of poly I, poly C, and a 1:1 mixture of the two in 0.1 M NaCl, 0.01 M NaOAc, pH 7.4 (86).

tions no reaction occurred in 10^{-4} M NaCl. In 0.01 M NaCl the process required 2 hr to attain equilibrium and in 0.1 M NaCl only several minutes (85).

Chamberlin and Patterson have extended these investigations to the deoxyribose analogues of rI and rC (87). All possible hybrid combinations have been observed, i.e., rI:rC; dI:dC; dI:rC; and rI:dC. If the mixing curves were obtained at neutral pH, 25°, and an ionic strength of 0.1 to 0.5, all four homopolymer pairs showed similar behavior. The mixing curves in each case were V-shaped with a sharp minimum at 50% rC or dC. For these conditions it is clear that only the 1:1 species is formed (88).

If the molarity of NaCl is increased to 0.6 or more, the dI:dC and dI:rC species can add a second strand of dI to form the three-stranded 2dI:dC or 2dI:rC species, respectively. This reaction does not appear to occur for the rI:rC or rI:dC bihelical forms (87).

Davies has concluded from an analysis of the X-ray diffraction pattern of fibers of the rI:rC complex that its structure is that of a doubly-stranded helix (88). The hydrogen bonding of the base pairs is between the C-6 carbonyl of hypoxanthine and the C-6 amino group of cytosine and between the N-1 of hypoxanthine and the N-1 of cytosine. The bonding is similar to that between guanine and cytosine in DNA, except for the absence of the third hydrogen bond involving the C-2 amino group of guanine. The strands are antiparallel.

Information about the conformation of the other homopolymer pairs is incomplete. Chamberlin and Patterson have reported that the X-ray diffraction patterns of fibers of rI:rC and rI:dC are essentially equivalent, but that dI:rC and dI:dC gave patterns differing radically from these and from each other (87). There is a definite implication that the latter two forms may have a different structure.

All of the bihelical complexes of dI or rI with dC or rC show the sharp thermally-induced helix → coil transition characteristic of structures of this kind. A somewhat unexpected feature is the pronounced differences in the thermal stability, as measured by the temperature (T_m) of the midpoint of the thermal transition. This increases in the order:

$$dI:rC < dI:dC < rI:dC < rI:rC$$

This order is preserved over a 100-fold range of electrolyte concentration.

E. The 1:1 Complex of Polyriboinosinic and Polyriboadenylic Acid

The hypoxanthine base of inosine is potentially capable of forming a hydrogen-bonded base pair with adenine. The combination of polyriboinosinic acid (poly I) and polyriboadenylic acid to form an ordered complex was first observed by Rich (89a). As in the cases of the other reactions discussed in this section, the process results in a major alteration in the ultraviolet absorption spectrum with a large increase in hypochromism. The decrease in absorbancy at 254 mμ provides a convenient index of the extent of reaction.

The process has many points of similarity to the poly A plus poly U interaction. The equilibrium mixing curves at ionic strengths of 0.1 or greater are V-shaped (Fig. 25) with a sharp minimum at a mole fraction (x_I) of poly I equal to 0.67 (90). At very short reaction times it has been reported that a transient minimum occurs at $x_I = 0.50$ (89a).

Although a rigorous analysis has not been made, by analogy with the poly A plus poly U case it appears possible that a complex of the rA:rI type is formed stoichiometrically at values of x_I less than 0.50 and that a transition to an rA:2rI complex occurs at higher values of x_I, attaining completion at $x_I = 0.67$. This model was presented by Rich (89a).

The rate of interaction is strongly dependent upon ionic strength (90). The velocity increases rapidly with increasing electrolyte concentration, with divalent cations exerting a disproportionate effect in enhancing the rate.

Like the other helical polynucleotides, the equimolar complex undergoes denaturation at extremes of pH and at elevated temperatures. As monitored by the increase in absorbancy at 254 mμ, both transitions are remarkably sharp and have the characteristics of cooperative processes.

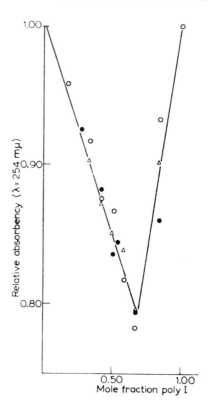

Fig. 25. Equilibrium mixing curves for poly A plus poly I at different ionic strengths in 0.001 *M* NaOAc, plus NaCl to indicated molarity (*2*): (○) 0.1 *M* NaCl, (●) 0.2 *M* NaCl, (△) 0.33 *M* NaCl.

In the case of the thermal transition the position of the midpoint is very dependent upon ionic strength and is displaced to higher temperatures with increasing ionic strength (*90*).

Rich has reported X-ray diffraction patterns for oriented fibers of the rA:rI complex (*89a*). The over-all appearance was similar to that of the B form of DNA. The detailed structure proposed by Rich on the basis of the diffraction pattern is somewhat similar to that of the rA:rU species. It consists of an antiparallel two-stranded helix with the purine bases in the core and the sugar–phosphate chains on the periphery. The bases, which are roughly perpendicular to the helical axis, are hydrogen-bonded, with bonding between the N-1 nitrogens of hypoxanthine and adenine and between the 6-amino group of adenine and the 6-carbonyl of hypoxanthine. The 1:1 complex reported by Rich may not be formed under some conditions (*89b*).

VII. DOUBLY-STRANDED HELICAL HOMOPOLYMERS

A. The Acid Form of Poly A

Poly A undergoes a dramatically sharp structural transition as the pH is lowered below a critical value in the vicinity of pH 6, the position of of the transition depending upon the ionic strength and temperature (*28,39,91*). In 0.1 *M* KCl the molecular transformation attains completion over a very narrow range of pH—less than 0.1 pH units (*28*).

The first indication of the structural change came from hydrogen ion titration curves of poly A (*39,91*), which showed a discontinuity at the critical pH (Fig. 26). The binding of hydrogen ions is strongly cooperative, the titration curve rising almost vertically until about half the bases are protonated. The pK of adenine is displaced by two pH units to the alkaline from its value (3.8) for the nucleotide monomer (*28*). An increase in temperature or ionic strength shifts the transition region to lower values of pH (*28b*). The fraction of adenine bases which must be

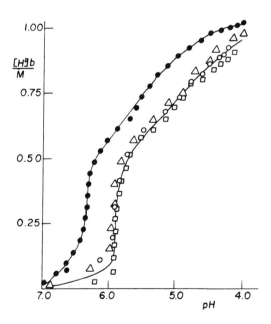

Fig. 26. Hydrogen ion titration curves of poly A at two ionic strengths at 26°. The ordinate is the number of hydrogen ions bound per AMP unit (37); (●) 0.01 *M* KCl, 0.53 × 10^{-3} *M* nucleotide; (○) 0.1 *M* KCl, 0.10 × 10^{-3} *M* nucleotide; (△) 0.1 *M* KCl, 0.27 × 10^{-3} *M* nucleotide; (□) 0.1 *M* KCl, 0.46 × 10^{-3} *M* nucleotide.

protonated for the acid form to be stable decreases with decreasing electrolyte concentration and may be as low as 0.2 (for 0.001 M KCl, 0°C) (*28b*).

The transition results in a major change in the ultraviolet absorption spectrum of poly A (*28,39,91*). A 20 to 25% drop in absorbancy occurs at 257 mμ, the position of the maximum for alkaline poly A (*28,91*). The maximum shifts to 252 mμ (Fig. 27). A parallel change in optical rotation occurs with a major increase in the positive component of the specific rotation (*30*).

The spectral and optical rotatory changes are accompanied by important alterations in the hydrodynamic properties. Fluorescence polarization studies, using an acriflavine conjugate of poly A, have indicated that an increase in rotational relaxation time occurs, suggesting that the product of the transition has a more rigid structure than alkaline poly A (*40*).

Some increase in molecular weight is always observed, the magnitude of the increase depending upon the polymer concentration as well as the temperature and ionic strength (*28*). The low-angle X-ray diffraction measurements of Luzzati and coworkers have indicated that acid poly A has a mass per unit length which is twice that of the alkaline form, being similar to that of DNA (*35*).

A detailed structure, which is based upon analysis of the X-ray dif-

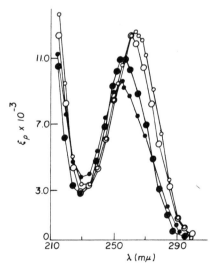

Fig. 27. The ultraviolet absorption spectrum of poly A at neutral and at acid pH, before and after formaldehyde treatment, which blocks the C-6 amino group and eliminates the structural transition (*37*). The solvent is 0.01 M KCl. (●) pH 7.0; (•) pH 5.0; (○) HCHO-treated, pH 7.0. (○) HCHO-treated, pH 5.0.

fraction patterns of oriented fibers (*92*), has been proposed for the acid form of poly A. This model consists of two helically wound strands with the sugar–phosphate backbones on the periphery and the bases in the core. The two strands are *parallel* rather than antiparallel. The repeat distance along the fiber axis is 3.8 Å. The bases, which are roughly normal to the fiber axis, are linked by two hydrogen bonds. One of these is between the C-6 amino group and the N-7 imidazole nitrogen. The other utilizes the second hydrogen of the amino group and one of the nonesterified oxygens of the phosphate. Such a structure would acquire additional stabilization from the electrostatic interaction between the positive charge of protonated adenine and the negatively charged phosphate group.

Whether the above structure is the only one present in acid poly A remains an open question. Certainly the occurrence of alternative structures has yet to be ruled out.

Despite the doubly-stranded character of the acid poly A helix, its hydrodynamic properties do not indicate a high degree of rigidity comparable to that of DNA. The intrinsic viscosity is not high and may, depending upon the conditions of formation, be either greater or smaller than that of the alkaline form (*36*). The interpretation of the hydrodynamic properties of acid poly A is complicated by the occurrence of random aggregation to a variable degree.

Fresco and Doty have found that the molecular weight of acid poly A increased with increasing concentration of the solution which was transformed. This was interpreted in terms of building up a series of homologous two-stranded helices of increasing length (*39*). However, the contribution of random aggregates was not assessed. The molecular weight dependence of sedimentation coefficient and intrinsic viscosity for a set of poly A preparations obtained in this way was:

$$[\eta] = KM^{0.91} \tag{9}$$

$$[S_{20}^0] = K'M^{0.36} \tag{10}$$

Because of the uncertain influence of aggregation, the values of the exponents should be regarded as only tentative. Subject to this reservation, their magnitude suggests a flexible and coiled configuration for acid poly A, which is however somewhat more rigid than the alkaline form.

Acid poly A undergoes the usual thermally-induced helix → coil transition (*93*). The transition is not gradual, as for alkaline poly A, but has the sharpness characteristic of the DNA-like class of polynucleotides (*93*). The midpoint, as determined from ultraviolet absorbancy, increases with decreasing pH and ionic strength (*93*). Both effects are predictable on electrostatic grounds, the effect of ionic strength probably stemming from a shift in the pK of adenine. Massoulié has found that the temperature of the transition midpoint varies linearly with pH for electrolyte concentrations of 0.03 to 0.5 (*72b*).

B. The Acid Form of Poly C

At a critical pH range in the vicinity of pH 6 poly C undergoes a structural transition to an acid form whose properties suggest that it is a two-stranded helix. Langridge and Rich have made a detailed X-ray diffraction study of fibers of the acid form of poly C and have proposed a structure (*94*). Their model is a double helix whose pitch is 18.65 Å and which has six residues per helical turn. The diameter is about 13.5 Å. The bases are in the core of the helix and are linked by three hydrogen bonds, two of which join the amino and carbonyl groups and the third the two N-1 nitrogens (Fig. 28). A distinctive feature of this structure is that only half the cytosine bases are protonated. It would therefore be expected that *complete* protonation of the bases would tend to destabilize the helix.

As in the case of the other helical polynucleotides, the optical rotation of acid poly C at wavelengths above 300 mμ is large and positive (*44,46*).

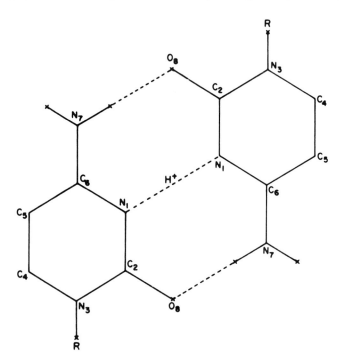

Fig. 28. Hydrogen bonding of cytosines in acid poly C (*94*).

The thermal profiles of ultraviolet absorbancy and optical rotation have the sharpness characteristic of two-stranded helical polynucleotides. The midpoint of the thermal transition increases with decreasing pH down to about pH 4 and subsequently decreases. This is undoubtedly related to the progressive protonation of the bases, as discussed above.

The hydrodynamic properties of acid poly C have yet to be examined in detail. The existing information indicates that a considerable molecular contraction accompanies the transition to the acid form. A decrease in intrinsic viscosity and radius of gyration and an increase in sedimentation coefficient, as compared with the alkaline form, have been reported (46).

The best model at present for the acid form of poly C is that of rather compact doubly-stranded helices, which have some points of similarity to acid poly A.

C. The Helical Form of Polyribouridylic Acid

As has already been discussed, polyribouridylic acid (poly U or rU) exists as a structureless coil at ordinary temperatures. However, Lipsett has found that at neutral pH poly U undergoes a transition to an organized structure at low temperatures (23). In 0.01 M MgCl$_2$ at pH 7.4 the absorbancy at 260 mμ drops by 30% upon cooling to 2°, the position of the maximum shifting from 260 to 258.5 mμ. The magnitude of the specific rotation at 589 mμ increases from 11 to 290°.

The helix → coil transition, as followed by absorbancy measurements, is of a sharpened character, like those of the other helical species described in this section. The midpoint of the thermal transition for the above conditions is 5.8°. The helical structure is also lost upon ionization of the uracil bases at alkaline pH.

While the molecular properties of poly U at low temperatures have the general features associated with an organized helical structure in polynucleotides, no definitive structural analysis has as yet been published. There is also no detailed information as to the hydrodynamic properties.

VIII. MULTIPLE-STRANDED HELICAL HOMOPOLYNUCLEOTIDES

A. Polyriboinosinic Acid

Rich has examined the X-ray diffraction pattern produced by oriented fibers of polyriboinosinic acid (poly I or rI). The poly I pattern, while of a helical type, is quite different in appearance from that of DNA or the other helical polynucleotides discussed earlier (95). Several possible helical structures were considered by Rich, who rejected one- and two-stranded models on steric grounds (*95*). However, a triply-stranded structure was found to be sterically feasible and consistent with the observed diffraction pattern.

The model proposed by Rich has three hydrogen bonds for each set of three bases. These link the N-1 nitrogen and the C-6 carbonyl of the hypoxanthine bases (Fig. 29). The helix is right-handed.

While the computation of a Fourier transform for this structure showed it to be consistent with the diffraction data, it was not possible to exclude other models entirely. In particular, a four-stranded structure could be constructed which had some degree of plausibility, although it was less consistent with the diffraction data than the three-stranded model.

The published information as to the solution properties of poly I is fragmentary. Rich has reported that the absorbancy at 250 mμ of a poly I solution at pH 7 decreased by 18% as the ionic strength was increased from 0.01 to 1.0 (*95*). A parallel ionic strength dependence was shown by the sedimentation coefficient, which increased from 5.9 Svedbergs in 0.2 M NaCl to 8.9 Svedbergs in 0.65 M NaCl. The thermal profile of absorbancy shows a rapid increase in a critical temperature range which depends upon the ionic strength. At pH 10, where the hypo-

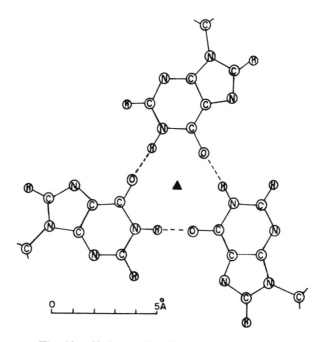

Fig. 29. Hydrogen bonding in poly I (*95*).

xanthine groups are ionized, the hypochromism largely disappears and the sedimentation coefficient drops by a considerable factor.

These observations are consistent with the presence of an organized form of poly I at neutral pH, which may be tentatively identified with the helical structure found in fibers.

B. Polyriboguanylic Acid

In view of the chemical similarity of polyriboguanylic acid (poly G or rG) to poly I, it is not surprising to find that it can also assume an organized helical structure. Fresco and Massoulié have found that, in 0.2 M NaCl at 25°, poly G has a large and positive specific optical rotation (100°), in contrast to the negative value (−26°) of the monomer, GMP, under the same conditions (96). The hypochromism is large, about 30% at 252 mμ (96). These data suggest strongly that poly G has an organized structure under these conditions.

The alkaline pK of poly G is shifted to 11.2 from a value for the monomer of 9.4. The titration curve has cooperative features, being markedly sharpened as compared with the monomer. The ionization of the guanine groups is accompanied by the loss of the hypochromism and dextrorotation, presumably indicating the breakup of the organized structure. A helix → coil transition also occurs at elevated temperatures.

Although definitive information is lacking, it has been proposed that the organized form of poly G has a multistranded helical structure, containing three, or possibly four, strands.

IX. TRIPLY-STRANDED HELICAL COMPLEXES OF DIFFERENT POLYNUCLEOTIDES

A. The rA:2rU Complex

At ionic strengths of 0.1 or greater the two-stranded rA:rU complex discussed in Section VI can, in the presence of excess poly U, add a second strand of poly U to form the three-stranded rA:2rU species. The reaction can be followed by the decrease in absorbancy at 260 mμ, which is equal to that resulting from the formation of the rA:rU species, so that the spectral mixing curve consists of two straight lines intersecting at a mole fraction of poly U equal to 0.67 (Fig. 19).

Rich (70,75) has proposed on stereochemical grounds that the second poly U strand lies in the deep helical groove of the rA:rU complex and is stabilized by hydrogen bonds between the C-6 carbonyl of uracil and the C-6 amino group of adenine and between the N-1 nitrogen of uracil and the N-7 nitrogen of adenine (Fig. 30). In Rich's model the C-6 carbonyls of the uracils of both poly U strands are involved in hydrogen bonding.

However, Miles has suggested an alternative structure (Fig. 30) in which the uracils in the two poly U strands are hydrogen bonded to the adenine in the poly A strand by *different* oxygen atoms (97). This structure is based upon infrared spectral evidence, which indicates that hydrogen bonding of both a C-6 and a C-2 carbonyl of uracil occurs in the rA:2rU species. In the structure of Miles both poly U strands are antiparallel to the poly A strand.

At 25°, for all solvents thus far investigated, the rA:2rU species is less stable than the rA:rU form, so that formation of the latter proceeds to completion (at a U:A ratio of 1:1) before addition of the second poly U strand. The rate of addition of the third strand also appears to be slower than the initial combination of poly A and poly U to form the two-stranded complex.

The pH and ionic strength dependence of the equilibria involving rA:rU and rA:2rU is complicated. Four different kinds of transition are possible, the thermal stability of each of which may be measured by its half transition temperature, T_m (72b). The four transitions and the designations of the corresponding transition midpoints are as follows:

$$rA:rU \rightleftharpoons rA + rU; T_{m_{2-1}}$$
$$rA:2rU \rightleftharpoons rA:rU + rU; T_{m_{3-2}}$$
$$rA:2rU \rightleftharpoons rA + 2rU; T_{m_{3-1}}$$
$$2(rA:rU) \rightleftharpoons rA + rA:2rU; T_{m_{2-3}}$$

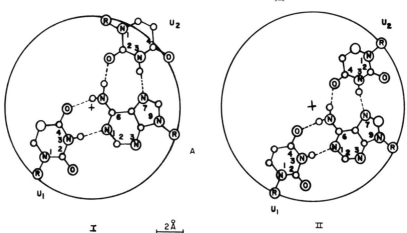

Fig. 30. Hydrogen bonding schemes for the models of rA:2rU proposed by Miles (I) and by Rich (II) (70,75,97). In this illustration, taken from reference 97, ring positions are given in the International system (see p. 66).

Although the ultraviolet absorbancies of rA:rU and rA:2rU are equivalent at 260 mμ, they differ at 280 mμ. Thus measurements of absorption spectra at different wavelengths can be used to determine the relative concentrations of the two species and to monitor the above equilibria (72b).

Using this approach, Massoulié (72b) has found that, at pH 7 and NaCl concentrations greater than 0.2, all four transition temperatures are well above 25°. At room temperature under these conditions the rA:rU species is formed stoichiometrically at rU to rA ratios up to 1.0. At higher rU to rA ratios the excess rU is incorporated stoichiometrically into rA:2rU. As the NaCl level is decreased, $T_{m_{3-2}}$ falls, becoming less than 25° for salt concentrations below 0.01. Under these conditions the rA:2rU complex is unstable at 25° and only the rA:rU species is formed. As the NaCl concentration is increased above 0.1, $T_{m_{2-3}}$ becomes lower than $T_{m_{3-1}}$, so that a temperature zone exists where rA:2rU is the only stable complex species.

In 0.1 M NaCl, both $T_{m_{2-1}}$ and $T_{m_{3-2}}$ decrease as the pH approaches the pK of the uracil groups of poly U ($\sim$9.7). Since $T_{m_{3-2}} < T_{m_{2-1}}$, first the rA:2rU and then the rA:rU species dissociate as the pH increases at 25°. At pH's above 10 both $T_{m_{2-1}}$ and $T_{m_{3-2}}$ become less than 25° and both complexes dissociate at this temperature.

In acid solution (below pH 6) the situation is further complicated by the structural transition of poly A to the acid form, which is also characterized by a transition temperature, T_{m_A}. T_{m_A} increases as the pH is lowered (72b). The transition temperatures of rA:rU and rA:2rU do not change as the pH is lowered until a pH is reached at which T_{m_A} is equal to those of the complex species. Below this pH $T_{m_{2-1}}$, $T_{m_{3-1}}$, and $T_{m_{2-3}}$ (but not $T_{m_{2-2}}$) fall with decreasing pH.

Massoulié has concluded that the acid dissociation of rA:rU and rA:2rU arises from the competitive formation of the acid form of poly A, rather than a destabilization of the complexes themselves (72b).

The hydrodynamic properties of the rA:2rU complex have yet to be examined. Fluorescence polarization measurements, using poly U conjugated with the fluorescent label acriflavine, have indicated that the rA:2rU species is appreciably more rigid in structure than the rA:rU form, with somewhat less freedom of internal rotation (40).

B. The rA:2rI Complex

In the presence of excess poly I the rA:rI species can add a second strand of poly I to form the three-stranded rA:2rI complex. In the structure which has been proposed for this species the C-6 amino group

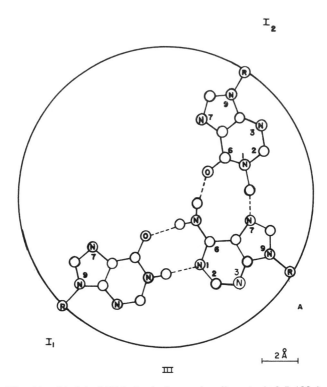

Fig. 31.   Model of Rich for hydrogen bonding at rA:2rI (*89a*).

of adenine is hydrogen bonded to the C-6 carbonyls of the hypoxanthine groups of the two poly I strands; the N-1 nitrogen of one hypoxanthine with the N-1 of adenine; and the N-1 of the second hypoxanthine with the imidazole N-7 nitrogen of adenine (*89a*) (Fig. 31).

X. COPOLYMERS OF RANDOM NUCLEOTIDE SEQUENCE

In contrast to DNA- and RNA-polymerase, the enzyme polynucleotide phosphorylase catalyzes the formation only of polyribonucleotides of random base sequence. These products are in some degree physical analogues of RNA, although of course devoid of biological specificity.

The most detailed physical information is available for copolymers whose base compositions are adenine–uracil (poly AU), adenine–hypoxanthine (poly AI), adenine–cytosine (poly AC), and hypoxanthine–cytosine (poly IC). The properties of each type have been examined as a function of the mole fractions of the constituent nucleotides (*98–101*).

In each case hydrodynamic and light-scattering data have indicated that, at neutral pH and ionic strength 0.1, the molecular extension is very much less than the contour length and presumably corresponds to an over-all randomly coiled configuration, resembling in this respect the alkaline forms of poly A and poly C. Nevertheless, although long-range order is probably absent, there is strong evidence for the presence of extensive short-range order. Thus all of the above copolymers have a high degree of hypochromism and a specific rotation at 589 mμ which is large and positive. In the case of poly AI the magnitude of the dextrorotation decreases with increasing hypoxanthine content (99).

At neutral pH and ionic strength 0.1 the thermal profiles of ultraviolet absorbancy show a gradual loss of hypochromism with increasing temperature and qualitatively resemble those of alkaline poly A and poly C. In all probability the broadness of the thermal transition can be attributed to the melting out of localized ordered regions of limited and variable extent.

In no case is definitive information available as to the conformation of the helical zones and, in particular, as to whether they are stabilized primarily by "vertical" interactions of stacked bases or whether "horizontal" interactions of hydrogen-bonded base pairs also make a contribution. In the cases of poly AU and poly IC, the latter factor may well be important in view of the known interaction of the two corresponding sets of homopolymers in each case to form doubly-stranded complexes. The antiparallel character of the latter makes it plausible that portions of the poly AU or poly IC chains might be bent into hairpin-shaped helical regions analogous in form to the helical duplexes of rA:rU or rI:rC.

It is interesting that, in the case of the adenine-containing copolymers, the capacity to interact with poly U is retained for quite high mole fractions, up to at least 0.5, of the second nucleotide (98,99,102). The stoichiometry of the process suggests that both two- and three-stranded complexes are formed.

XI. RNA

A. Molecular Structure of RNA

We shall confine attention here to the single-stranded varieties of RNA. The few examples of natural RNA's which are definitely known to be two-stranded, such as that of reovirus, appear to be essentially equivalent in physical properties to DNA of comparable molecular weight and do not require special comment here.

Definitive physical information is available for three different classes of natural RNA, which should be differentiated, as there exist no grounds for assuming their molecular organizations to be equivalent. *Soluble* RNA (*s*RNA), which occurs in the free state in the cytoplasm of all living cells, has an essential function in protein synthesis as a form of adaptor to which amino acids must be temporarily attached prior to incorporation into polypeptide. *Ribosomal* RNA, whose biological function is uncertain, is found in ribosomes and accounts for about half their mass. *Viral* RNA replaces DNA as the basic carrier of genetic information for a number of viruses, including tobacco mosaic virus. To the above should of course be added *messenger* RNA, the actual directive agent in protein synthesis. However, since detailed information as to its physical properties is lacking, it will not be discussed here.

In no case is the three-dimensional structure known in detail, although the nucleotide sequence of several forms of *s*RNA have been established. Since the regularity of base composition characteristic of DNA is not present, it is not to be expected that the entire molecule can be incorporated into a perfect two-stranded helix of the DNA type. On the other hand there is abundant evidence that all of the natural RNA's thus far examined have a high degree of helical content of some kind.

Most of the speculation as to the structure of RNA has been guided by the view that the helical regions present are of the DNA type and that the same base pairings occur (with uracil replacing thymine). There are of course no grounds for excluding other types of base pairing in particular instances. Since RNA is single stranded, helical regions of this kind can only arise by a bending back of the strand upon itself in a hairpinlike arrangement (Fig. 32). While the X-ray diffraction evidence is inconclusive (*103,104*), the existing information is not inconsistent with this kind of model.

The detailed models of this kind which have been proposed fall into two main categories, which are not mutually exclusive.

(1) There exist within the same RNA strand extensive sequences of nucleotides which are mutually complementary and which can form one or more hairpin-shaped helical regions of the DNA type, the remainder of the strand being essentially structureless (Fig. 32a).

(2) There are no extensive regions of complementary base sequence. However, it is still sterically feasible to form a doubly-stranded helical structure from which mismatched nucleotides are excluded as external loops (Fig. 32b), thereby permitting the incorporation of most of the molecule in a DNA-like helix (*105*).

In addition to the above there exists a minority viewpoint that the structure of RNA may not depend upon base pairing at all, but is instead analogous to the alkaline forms of poly A and poly C (*106*).

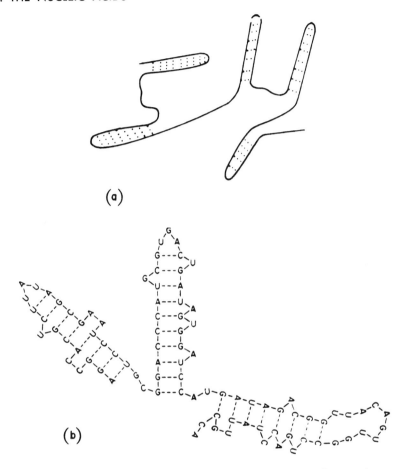

Fig. 32. Two models for the structure of RNA. (a) Mutually complementary sequences. (b) Exclusion of mismatched bases.

None of the above models is explicit enough to be of much help in interpreting the solution properties of RNA. It is entirely possible that all three are important, their relative contribution varying for different RNA's.

B. Viral RNA

The discussion of viral RNA will be confined to that isolated from tobacco mosaic virus (TMV), for which the most extensive information is available. All of the RNA of TMV (molecular weight 40×10^6) occurs as a single strand of molecular weight close to 2.0×10^6 (*107*).

This has been isolated in a purified, protein-free state and shown to retain infectious properties.

Boedtker has found a radius of gyration of about 300 Å at an ionic strength of 0.06 and pH 8.5 from light-scattering measurements (*107*). A rodlike model for the molecule can certainly be rejected since this radius of gyration corresponds to an end-to-end separation only about $\frac{1}{25}$th of that predicted for a completely extended rigid helix.

Moreover, if the intrinsic viscosity and the sedimentation coefficient are combined in the Flory–Mandelkern equation (*108*) to compute a molecular weight upon the assumption of a randomly coiled configuration, a value of 1.8×10^6 is obtained. This is close enough to the actual figure to indicate that the randomly coiled model provides a reasonably self-consistent picture of the gross over-all configuration, at least at high ionic strengths.

The reduced specific viscosity was found to be ionic-strength dependent, a twofold decrease accompanying an increase in ionic strength from zero to 0.2 at neutral pH (Table IIIB). This result certainly indicates some degree of flexibility which permits extension under electrostatic stress, although the effect is much less than for a structureless polynucleotide such as poly U.

It was also observed that the viscosity of this RNA was temperature-dependent, a 50% increase occurring reversibly between 6 and 25°. In harmony with this finding, light scattering revealed a definite expansion of the molecule with temperature.

The absorbancy at 260 mμ is also temperature-dependent, increasing

TABLE IIIB
Viscosity of TMV RNA at 5° (*107*)

Ionic strength	Solvent	pH	c, g/100 ml	η_{sp}/c
0.20	Phosphate	7.0	0.027	0.82
0.08	Phosphate	7.5	0.0935	0.81
	Plus versene		0.0537	0.77
0.06	Phosphate	8.5	0.0745	0.82
			0.0435	0.705
			0.0396	0.725
0.06	Versene	7.5	0.1235	0.82
			0.0740	0.775
			0.0247	0.70
0.04	Phosphate	7.0	0.067	1.10
			0.032	1.19
0	Water	6.0	0.023	1.60

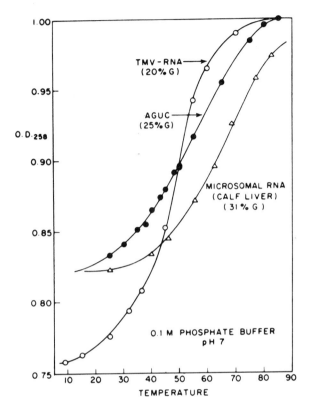

Fig. 33. Temperature dependence of ultraviolet absorbancy for several RNA's of varying guanine content in 0.1 M phosphate, pH 7 (64). AGUC, which was prepared by the action of polynucleotide phosphorylase, is a biosynthetic poly-ribonucleotide containing all four bases.

with increasing temperature (Fig. 33). However, the transition does not have the sharpness of that of DNA, but occurs gradually over a wide range of temperature.

The most self-consistent picture of the RNA from TMV which emerges from the above observations is that of a molecule containing helical regions of limited extent, whose over-all shape is that of a coil of moderate flexibility.

C. Soluble RNA

Soluble RNA consists of a collection of molecular species of different biological specificity with respect to their capacity to accept amino acids. All share the terminal trinucleotide sequence -CCA. Since complete physical information is not available for a single purified species, it will be necessary for many purposes to discuss *s*RNA as a single entity, bearing in mind that the observed parameters may be only averages for species differing perceptibly in properties. Some degree of justification for this somewhat arbitrary procedure is provided by the similar molecular weight and parallel biological function of all species. The physical parameters which have been reported for *s*RNA are summarized in Table IV.

On an average, it appears that *s*RNA from all sources centers around a molecular weight of about 25,000 to 28,000 with a sedimentation coefficient ($S_{20,w}^0$) of approximately 4 Svedbergs. The variations in the latter value and that of the intrinsic viscosity may possibly be attributed to differences in solution ionic strength or to minor degradation.

With regard to the possible molecular conformations of *s*RNA, the cal-

TABLE IV
Hydrodynamic Parameters of *s*RNA from Different Sources[a]

Source	Molecular weight	Sedimentation coefficient	Intrinsic viscosity	Diffusion coefficient $\times 10^7$	Equivalent ellipsoid axial ratio	Partial specific volume
Calf liver[b]	27,500	3.85				
E. coli[b]	28,000	3.92				
Yeast[b]	28,000	4.05				
Trout liver[b]	—	3.83				
Pea seedlings[b]	—	3.76				
Tetrahymena Pyr.[b]	—	3.92				
Blowfly larvae[b]	27,000	3.90				
E. coli[c]	24,500					
Yeast[d]	25,000	—				
E. coli[e]	25,500	4.0	0.075	7.84		0.55
Yeast[f]	30,000	4.0				
Yeast[g]	26,400	4.0	0.060		7	0.531
Yeast[h]	23,000	4.0	0.050			
Rabbit liver[i]	23,000		0.15		6	0.47
Wheat germ[j]	28,200	3.98	0.071	~6.5	5	0.53

[a] Room temperature, pH near neutrality, ionic strength ~0.1.

[b] G. L. Brown, Z. Kosinski, and C. Carr, in *Acides Ribonucleiques et Polyphosphates*, C.N.R.S., 1961, p. 183.

[c] G. L. Brown and G. Zubay, *J. Mol. Biol.*, **2**, 287 (1960).

[d] M. Litt and V. M. Ingram, *Biochemistry*, **3**, 560 (1964).

[e] A. Tissières, *J. Mol. Biol.*, **1**, 365 (1959).

[f] J. T. Penniston and P. Doty, *Biopolymers*, **1**, 145 (1963).

[g] T. Lindahl, D. D. Henley, and J. R. Fresco, *J. Amer. Chem. Soc.*, **87**, 4961 (1965).

[h] S. Osawa, *Biochim. Biophys. Acta*, **43**, 110 (1960).

[i] S. W. Luborsky and G. Cantoni, *Biochim. Biophys. Acta*, **61**, 481 (1962).

[j] C. McKay and K. Oikawa, *Biochemistry*, **5**, 213 (1966).

culation of the axial ratio of the hydrodynamically equivalent prolate ellipsoid of revolution is at present the most experimentally accessible means of providing information.   In neutral salt, the computed axial ratio of this theoretical counterpart of sRNA appears to be about 5–7.   It should be stressed that a *literal* transposition of this value for the hydrodynamic equivalent ellipsoid to an actual molecular shape of sRNA may be erroneous.   However, all estimations certainly point to a rather asymmetric hydrodynamic model and it would be surprising if the true molecular conformation was not of this type.

The nucleotide sequence of alanyl sRNA is completely known at present (*109*).   The three-dimensional geometrical arrangement of the polynucleotide strand, even if it is assumed that all species have a similar configuration, remains uncertain, although a "clover-leaf" model permits maximum base pairing and has been widely accepted.   While an X-ray diffraction study of several years ago seemed to indicate that the prevalent structure was that of a hairpin-shaped helix of the DNA type, both results and conclusion have subsequently been retracted (*103,104*).

Henley, Lindahl, and Fresco have examined the intrinsic viscosity and sedimentation coefficient of unfractionated sRNA from yeast at a series of temperatures (*110*).   These parameters can be combined in the equation of Scheraga and Mandelkern to yield the shape parameter, β, introduced by these authors (*111*).

$$\beta = \frac{S^0[\eta]^{1/3}\eta_0 N_0}{M^{2/3}(1 - \bar{V}\rho_0)} \tag{11}$$

where S^0 = sedimentation coefficient at infinite dilution, $[\eta]$ = intrinsic viscosity, η_0 = solvent viscosity, N_0 = Avogadro's number, M = molecular weight, $\bar{V}$ = partial specific volume, and ρ_0 = solvent density.

The parameter β gives an index of the shape of the equivalent unhydrated and impermeable ellipsoid of revolution which simulates the hydrodynamic properties of the actual system.   For prolate ellipsoids β ranges in value from 2.12×10^6 for an axial ratio of unity to 3.2×10^6 for an axial ratio of 100.

Below 20° in 0.1 M NaCl the computed value of β was anomalously large, suggesting that some degree of association occurred.   The value at 20°, 2.33×10^6, would formally correspond to an axial ratio of 7.   To the extent that the ellipsoidal model has some relation to reality, this result suggests a moderately asymmetric shape for sRNA.

The thermal profile of ultraviolet absorbancy at 270 mμ shows a gradual increase between 20 and 100° without the sharpening characteristic of the DNA-like class of polynucleotides.   The intrinsic viscosity increases and the sedimentation coefficient decreases over the same temperature range in rough synchrony with the absorbancy.   The value of β decreases appreciably between 20 and 30° and is nearly constant thereafter (Fig. 34).

Because we lack a definite model for the molecular form of native sRNA, it is not feasible to fit these observations into an explicit picture of

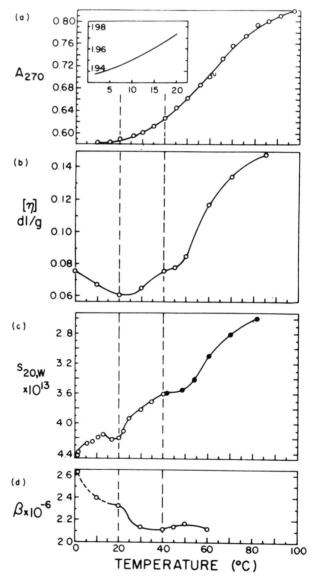

Fig. 34.   Temperature dependence of some physical parameters of yeast sRNA in 0.2 M NaCl, 0.01 M phosphate, pH 6.85 (110).

its thermal denaturation. The existing information is roughly in accord with expectations for the thermal melting of a set of limited helical regions, accompanied by an expansion of the molecule from an initially compact shape to a more random and open conformation, somewhat like the denaturation of a globular protein. Fluorescence polarization measurements upon an acriflavine conjugate of sRNA have indicated that this process results in an increase in flexibility (112).

Fresco, Klotz, and Richards (113) have examined the thermal profiles of absorbancy over a range of wavelengths for homogeneous preparations of yeast sRNA specific for alanine, valine, and tyrosine. In each case the midpoint of the transition was a function of wavelength. The profiles of absorbancy at 280 mμ were biphasic. These results were interpreted in terms of the independent melting of at least two helical phases which melt individually by a cooperative mechanism. These are postulated to differ in guanine–cytosine content, thereby accounting for both the different melting points and the different wavelength dependence of hypochromism. The absence of biphasic character for unfractionated preparations of sRNA can presumably be attributed to the loss of detail resulting from the summation of a set of different profiles.

In the presence of Mg^{2+} ion the properties of sRNA are appreciably altered. The effect is accentuated at low ionic strengths, where the binding of Mg^{2+} by phosphate groups is most pronounced. The midpoint of the thermal transition is displaced to higher temperatures. The intrinsic viscosity at 25° is lowered. At finite concentrations there is some degree of molecular association (112). These effects can undoubtedly be largely attributed to the neutralization of the negative electrostatic field of the sRNA molecule through binding of the oppositely charged Mg^{2+} ions.

The pronounced effect of magnesium ion in raising the T_m of the absorbance thermal profile of sRNA has its parallel in the rotatory dispersion curve (106). However, an interesting disparity between the percentage change in absorbance and rotation is revealed (106). Thus using the absorbance at 260 mμ as an indicator of secondary structure, the transition begins at 48° while the change in the Cotton effect begins almost immediately as the temperature is raised from 25°. In this case it is clear that both techniques exhibit different degrees of sensitivity in registering changes in conformation, optical rotatory dispersion being the more effective of the two.

An insight into the nature of the bonds which maintain the conformation of sRNA is afforded by the effect of ethylene glycol on the absorbance at 260 mμ and on the optical rotatory dispersion (106). The loss in structure as measured by both parameters is gradual at first. However, at concentrations greater than 50% the loss in structure becomes quite sharp with increasing ethylene glycol. A deviation between the two curves arises at higher levels of disorganization, optical rotatory dispersion indicating a greater loss of structure than absorbance. An exact explanation for the divergence between the two techniques is difficult at the present time.

Since the helical conformation of this polynucleotide is destroyed by ethylene glycol, a solvent believed to disrupt hydrophobic bonding, the inference is that hydrophobic bonding also plays an important role in stabilizing the structure of sRNA. A direct experimental estimate of the importance of hydrogen bonding has been made by the use of formaldehyde (106). Since formaldehyde reacts with the nucleotide amino groups, effectively blocking hydrogen bond formation at these sites, the amount of structural disorganization caused by this treatment is a measure of the stabilizing contribution of hydrogen bonding. The thermal profile of the optical parameters of formaldehyde-treated sRNA shows a relatively gradual transition somewhat like poly C, with or without formaldehyde (106). This result suggests that some sort of structure may be maintained in the absence of hydrogen bonding. The importance of hydrophobic bonding in native sRNA is strongly implied.

The ease of reversal of the structural transformations of sRNA depends upon conditions. Adams, Lindahl, and Fresco (106b) have reported that leucine sRNA may be converted to a metastable "denatured" state by heating at 60° in 0.01 M NaCl in the absence of Mg^{2+} and subsequent rapid cooling to 25°. The denatured species, which cannot accept leucine, may be reconverted to the native form by reheating briefly at 60° in the presence of 10^{-2} M Mg^{2+}.

The native and denatured species are of equivalent molecular weight and are similar with respect to hypochromism and optical rotation, but differ significantly in hydrodynamic properties. The denatured form has a higher intrinsic viscosity and a lower sedimentation coefficient. Since, in view of the similarity in hypochromism and optical rotation, the secondary structures must be similar, it has been postulated that the two forms differ in tertiary structure. This requires that nucleotide interactions other than those involved in base pairing must contribute to the over-all molecular structure.

D. Ribosomal RNA

The ribosomal RNA's of a number of species appear to have similar properties. The most detailed information is available for that from *E. coli*. This consists of two molecular species, which have been designated in terms of their sedimentation coefficients at infinite dilution as the 16S and 23S species (*114*).   While much early work appeared to suggest a structure based on discrete subunits which could be dissociated by thermal treatment (*115*), subsequent studies have provided strong evidence that this is not the case and that, in the absence of enzymic degradation, both forms consist of uninterrupted polynucleotide strands (*114*).   The molecular weights of the 16S and 23S species are close to 5×10^5 and 10×10^5, respectively, as determined from the Flory–Mandelkern relationship.

The molecular weight of the 23S species, as estimated from the Flory–Mandelkern equation on the assumption of a randomly coiled configuration, agrees with the value computed directly from sedimentation equilibrium data (*114*).

The randomly coiled model appears therefore to provide an internally consistent picture of the hydrodynamic properties.   This refers of course to the over-all shape and is not inconsistent with the presence of extensive regions of localized order.

That such ordered helical regions exist is strongly indicated by the high degree of hypochromism, which exhibits temperature-dependence; by the large positive component of the optical rotation at wavelengths above 300 mμ; by the character of the optical rotatory dispersion; and by the magnitude of the circular dichroism.   As in the case of sRNA, it is not as yet possible to specify any details of the helical conformation.

The intrinsic viscosity of unfractionated ribosomal RNA in 0.05 M tris, pH 7.3, has been reported to vary between 0.33 and 0.42 dl/g depending

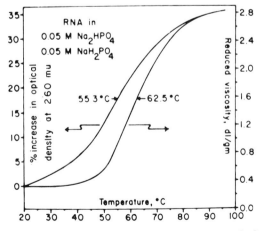

Fig. 35.   Temperature dependence of absorbancy at 260 mμ and of intrinsic viscosity for unfractionated ribosomal RNA from *E. coli* (*114*).

on the prior treatment of the sample. The value of the Huggins constant k in the equation

$$\eta_{\mathrm{sp}}/c = [\eta] + k[\eta]^2 c \qquad (12)$$

has been found to fall between 0.6 and 1.1.

The thermally-induced helix $\rightarrow$ coil transition has a qualitative resemblance to those of the other RNA's discussed in this section. The thermal profile of ultraviolet absorbancy is relatively broad. A pronounced rise in viscosity occurs in the melting region (Fig. 35).

In the case of calf liver microsomal RNA, the molecular weight dependence of the intrinsic viscosity and sedimentation coefficient for a series of degraded preparations has been reported to be (*115*)

$$[\eta] = K M^{0.53} \qquad (13)$$

$$S = K' M^{0.49} \qquad (14)$$

Both exponents are close to the values predicted (0.5) for a system of random coils in an ideal solvent.

REFERENCES

1. E. Chargaff and J. N. Davidson, *The Nucleic Acids,* Academic, New York, 1955.
2. R. F. Steiner and R. F. Beers, *Polynucleotides* Elsevier, Amsterdam, 1961.
3. A. R. Peacocke, "Structure and Physical Chemistry of Nucleic Acids," in *Progr. Biophys.* **10**, 55 (1959).
4. D. O. Jordan, *The Chemistry of the Nucleic Acids,* Butterworth, London, 1960.
5. A. M. Michelson, *The Chemistry of Nucleosides and Nucleotides,* Academic, New York, 1963.
6. R. L. Sinsheimer, in Ref. 1, Vol. 3, p. 187.
7. T. L. V. Ulbricht, in *Comprehensive Biochemistry,* Vol. 8 (M. Florkin and E. Stotz, eds.), Elsevier, Amsterdam, 1963, p. 158.
8. D. B. Dunn and J. D. Smith, *Nature,* **175**, 336 (1955).
9. J. W. Littlefield and D. B. Dunn, *Nature,* **181**, 254 (1958).
10. H. Amos and M. Korn, *Biochim. Biophys. Acta,* **34**, 286 (1959).
11. D. B. Dunn, *Biochim. Biophys. Acta,* **34**, 286 (1959).
12. M. Adler, B. Weissman, and A. B. Gutman, *J. Biol. Chem.,* **230**, 717 (1958).
13. J. D. Smith and D. B. Dunn, *Biochem. J.,* **72**, 294 (1959).
14. J. D. Watson and F. H. C. Crick, *Nature,* **171**, 737 (1953).
15. R. Langridge, H. R. Wilson, C. W. Hooper, M. H. F. Wilkins, and L. D. Hamilton, *J. Mol. Biol.,* **2**, 19, 38 (1960).
16. J. Donohue, *Proc. Nat. Acad. Sci. U. S.,* **42**, 60 (1954).
17. L. Pauling and R. Corey, *Arch. Biochem. Biophys.,* **65**, 164 (1956).
18. D. M. Crothers and B. H. Zimm, *J. Mol. Biol.,* **9**, 1 (1964).
19. A. Rich, in *The Chemical Basis of Heredity* (W. D. McElroy and B. Glass, eds.), Johns Hopkins Press, Baltimore, 1957, p. 557.
20. R. C. Warner, *J. Biol. Chem.,* **229**, 711 (1957).
21. J. R. Fresco, *Trans. N. Y. Acad. Sci.,* **21**, 653 (1959).
22. P. O. P. Ts'o, G. K. Helmkamp, and C. Sander, *Biochim. Biophys. Acta,* **55**, 584 (1962).

23. M. N. Lipsett, *Proc. Natl. Acad. Sci. U. S.,* 46, 445 (1960).

24. E. G. Richards, C. P. Flessel, and J. R. Fresco, *Biopolymers,* 1, 431 (1963).

25. R. C. Warner and E. Breslow, *Proc. Intern. Congr. Biochem. 4th, Vienna,* 1958, Vol. 9, p. 157.

26. L. Mandelkern and P. J. Flory, *J. Chem. Phys.,* 20, 212 (1952).

27. R. A. Cox, *J. Polym. Sci.,* 47, 441 (1960).

28a. R. F. Steiner and R. F. Beers, *Biochim. Biophys. Acta,* 32, 166 (1959).

28b. D. N. Holcomb and S. N. Timasheff, *Biopolymers,* 6, 513 (1968).

29. Ref. 2, p. 193.

30. J. Fresco, *Trans. N. Y. Acad. Sci.,* 21, 653 (1959).

31. I. Tinoco, *J. Amer. Chem. Soc.,* 82, 4785 (1960); 83, 5047 (1961).

32. B. H. Zimm and N. R. Kallenbach, *Ann. Rev. Phys. Chem.,* 13, 171 (1962).

33. J. Brahms, *J. Mol. Biol.,* 11, 785 (1965).

34. J. Brahms and W. F. H. M. Mommaerts, *J. Mol. Biol.,* 10, 73 (1964).

35. V. Luzzati, A. Mathis, F. Masson, and J. Witz, *J. Mol. Biol.,* 10, 28 (1964).

36. D. N. Holcomb and I. Tinoco, *Biopolymers,* 3, 121 (1965).

37. R. Steiner and R. Beers, *Biochim. Biophys. Acta,* 26, 336 (1957).

38. L. Treloar, *Proc. Phys. Soc.* (London), 55, 345 (1943).

39. J. Fresco and P. Doty, *J. Amer. Chem. Soc.,* 79, 3928 (1957).

40a. D. B. S. Millar and R. F. Steiner, *Biochim. Biophys. Acta,* 102, 571 (1965).

40b. H. Eisenberg and G. Felsenfeld, *J. Mol. Biol.,* 30, 17 (1967).

41. G. Weber, *Biochem. J.,* 51, 145, 155 (1952).

42. R. F. Steiner and H. Edelhoch, *Chem. Rev.,* 62, 457 (1962).

43. P. O. P. Ts'o, G. K. Helmkamp, and C. Sander, *Biochim. Biophys. Acta,* 55, 584 (1962).

44. G. D. Fasman, C. Lindblow, and L. Grossman, *Biochemistry,* 3, 1015 (1965).

45. J. Brahms, *J. Amer. Chem. Soc.,* 85, 3298 (1963).

46. E. O. Akinrimisi, C. Sander, and P. O. P. Ts'o, *Biochemistry,* 2, 340 (1963).

47. R. Langridge, W. Seeds, H. Wilson, C. Hooper, M. Wilkins, and L. Hamilton, *J. Biophys. Biochem. Cytol.,* 3, 767 (1957).

48. M. Feughelman, R. Langridge, W. Seeds, A. Stokes, H. Wilson, C. Hooper, M. H. F. Wilkins, R. Barclay, and L. Hamilton, *Nature,* 175, 834 (1955).

49. M. H. F. Wilkins, in *Comprehensive Biochemistry,* Vol. 8 (M. Florkin and E. Stotz, eds.), Elsevier, Amsterdam, 1963, p. 270.

50. D. Marvin, M. Spencer, M. H. F. Wilkins, and L. Hamilon, *Nature,* 182, 387 (1958).

51. Ref. 2, p. 202.

52. P. Doty, B. McGill, and S. Rice, *Proc. Nat. Acad. Sci. U. S.,* 44, 432 (1958).

53. C. L. Sadron, in Ref. 1, Vol. 3, p. 1.

54. E. Geiduschek and A. Holtzer, *Advan. Biol. Med. Phys.,* 6, 431 (1958).

55. G. Porod, *Monatsh. Chem.,* 80, 251 (1949).

56. B. Conway and J. Butler, *J. Polym. Sci.,* 12, 199 (1954).

57. P. Doty, *J. Cell. Comp. Physiol.,* 49, Suppl. 1, 27 (1957).

58. J. Sturtevant, S. Rice, and E. Geiduschek, *Discussions Faraday Soc.,* 25, 138 (1958).

59. D. Jordan, in Ref. 1, Vol. 1, p. 447.

60. R. Cox and A. Peacocke, *J. Chem. Soc.,* 1956, 2499.

61a. R. Cox and A. Peacocke, *J. Chem. Soc.,* 1957, 4724.

61b. C. Zimmer, G. Luck, H. Venner, and B. H. Zimm, *Biopolymers,* 6, 563 (1968).

62. L. Cavalieri and B. Rosenberg, *J. Amer. Chem. Soc.,* 79, 5352 (1957).

63. F. W. Studier, *J. Mol. Biol.,* 11, 373 (1965).

64. P. Doty, H. Boedtker, J. Fresco, R. Haselkorn, and M. Litt, *Proc. Natl. Acad. Sci. U. S.,* **45**, 482 (1959).

65. S. A. Rice and P. Doty, *J. Amer. Chem. Soc.,* **79**, 3937 (1957).

66. M. Meselson and F. Stahl, *Proc. Nat. Acad. Sci. U. S.,* **44**, 671 (1958).

67a. P. Doty, J. Marmur, J. Eigner, and C. Schildkraut, *Proc. Nat. Acad. Sci. U. S.,* **46**, 461 (1960).

67b. J. A. Subirana and P. Doty, *Biopolymers,* **4**, 171 (1966).

67c. J. G. Wetmur and N. Davidson, *J. Mol. Biol.,* **31**, 349 (1968).

68a. A. Peacocke, *J. Mol. Biol.,* **5**, 564 (1962).

68b. J. A. Harpst, A. I. Krasna, and B. H. Zimm, *Biopolymers,* **6**, 595 (1968).

69. R. Warner, *Ann. N. Y. Acad. Sci.,* **69**, 314 (1957).

70. G. Felsenfeld and A. Rich, *Biochim. Biophys. Acta,* **26**, 457 (1957).

71. R. F. Steiner and R. F. Beers, *Biochim. Biophys. Acta,* **33**, 470 (1959).

72a. C. L. Stevens and G. Felsenfeld, *Biopolymers,* **2**, 293 (1964).

72b. J. Massoulié, *Eur. J. Biochem.,* **3**, 428, 439 (1968).

73. J. Fresco, in *Informational Macromolecules,* Academic, New York, 1963, p. 121.

74. H. T. Miles and J. Frazier, *Biochem. Biophys. Res. Commun.,* **14**, 21 (1964).

75. A. Rich and D. Davies, *J. Amer. Chem. Soc.,* **78**, 3548 (1956).

76. A. Rich, in *The Chemical Basis of Heredity* (W. McElroy and B. Glass, eds.), Johns Hopkins Press, Baltimore, 1957, p. 557.

77. L. Katz, K. Tomita, and A. Rich, *J. Mol. Biol.,* **13**, 340 (1965).

78. P. Ross and J. Sturtevant, *Proc. Nat. Acad. Sci. U. S.,* **46**, 1360 (1960).

79. R. F. Steiner and C. Kitzinger, *Nature,* **194**, 1172 (1962).

80. M. Rawitscher, P. Ross, and J. Sturtevant, *J. Amer. Chem. Soc.,* **85**, 1915 (1963).

81. H. K. Schachman, J. Adler, C. M. Radding, I. R. Lehman, and A. Kornberg, *J. Biol. Chem.,* **235**, 3242 (1960).

82. D. R. Davies and R. L. Baldwin, *J. Mol. Biol.,* **6**, 251 (1963).

83. R. B. Inman, and R. L. Baldwin, *J. Mol. Biol.,* **5**, 172, 185 (1962).

84. H. Spatz and R. L. Baldwin, *J. Mol. Biol.,* **11**, 213 (1965).

85. R. B. Inman and R. L. Baldwin, *J. Mol. Biol.,* **5**, 173, 185 (1962).

86. D. R. Davies and A. Rich, *J. Amer. Chem. Soc.,* **80**, 1003 (1958).

87. M. J. Chamberlin and D. L. Patterson, *J. Mol. Biol.,* **12**, 410 (1965).

88. D. R. Davies, *Nature,* **186**, 1030 (1960).

89a. A. Rich, *Nature,* **181**, 521 (1958).

89b. P. B. Sigler, D. R. Davies, and H. T. Miles, *J. Mol. Biol.,* **5**, 709 (1962).

90. Ref. 2, p. 225.

91. R. F. Beers and R. F. Steiner, *Nature,* **179**, 1076 (1957).

92. A. Rich, D. R. Davies, F. H. C. Crick, and J. D. Watson, *J. Mol. Biol.,* **3**, 71 (1961)

93. P. O. P. Ts'o, G. K. Helmkamp, and C. Sander, *Proc. Nat. Acad. Sci. U. S.,* **48**, 686 (1962)

94. R. Langridge and A. Rich, *Nature,* **198**, 725 (1963).

95. A. Rich, *Biochim. Biophys. Acta,* **29**, 502 (1958).

96. J. Fresco and J. Massoulié, *J. Amer. Chem. Soc.,* **85**, 1352 (1963).

97. H. T. Miles, *Proc. Nat. Acad. Sci. U. S.,* **51**, 1104 (1964).

98. R. F. Steiner, *J. Biol. Chem.,* **235**, 2946 (1960).

99. R. F. Steiner, *J. Biol. Chem.,* **236**, 842 (1961).

100. R. F. Steiner, *J. Biol. Chem.,* **236**, 3037 (1961).

101. R. A. Cox, *Biochim. Biophys. Acta,* **72**, 203 (1963).

102. R. F. Steiner, *Ann. N. Y. Acad. Sci.,* **81**, 742 (1959).

103. M. Spencer, W. Fuller, M. H. F. Wilkins, and G. L. Brown, *Nature,* **194,** 1014 (1962).
104. M. Spencer and F. Poole, *J. Mol. Biol.,* **11,** 314 (1965).
105. J. R. Fresco, B. M. Alberts, and P. Doty, *Nature,* **188,** 98 (1960).
106a. G. Fasman, C. Lindblow, and E. Seaman, *J. Mol. Biol.,* **12,** 630 (1965).
106b. A. Adams, T. Lindahl, and J. R. Fresco, *Proc. Nat. Acad. Sci. U. S.,* **57,** 1684 (1967).
107. H. Boedtker, *Biochim. Biophys. Acta,* **32,** 519 (1959).
108. L. Mandelkern and P. J. Flory, *J. Chem. Phys.,* **20,** 212 (1952).
109. R. W. Holley, J. Apgar, G. A. Everett, J. T. Madison, M. Marquesee, S. H. Merrell, J. R. Penswick, and A. Zamir, *Science,* **147,** 1462 (1965).
110. D. Henley, T. Lindahl, and J. Fresco, *Proc. Nat. Acad. Sci. U. S.,* **55,** 191 (1966).
111. H. A. Scheraga and L. Mandelkern, *J. Amer. Chem. Soc.,* **75,** 179 (1953).
112. D. B. S. Millar and R. F. Steiner, *Biochemistry,* **5,** 2289 (1966).
113. J. R. Fresco, L. C. Klotz, and E. G. Richards, *Cold Spring Harbor Symp. Quant. Biol.,* **28,** 83 (1963).
114. W. M. Stanley, Jr., and R. M. Beck, *Biochemistry,* **4,** 1302 (1965).
115. B. Hall and P. Doty, in *Microsomal Particles and Protein Synthesis,* Washington Academy of Science, 1958.

CHAPTER 3

PHYSICAL CHEMISTRY OF ACIDIC POLYSACCHARIDES

Frederick A. Bettelheim

CHEMISTRY DEPARTMENT ADELPHI UNIVERSITY
GARDEN CITY, NEW YORK

I. INTRODUCTION

Among the biological polyelectrolytes, the acidic polysaccharides occupy a special position from many points of view. The polyanions are characterized (1) by their polysaccharidic backbones, and (2) by the

131

special anionic groups (mainly —COO⁻ and —SO₃⁻) and their spatial charge distribution along the backbone of the polymer. Most of the physical-chemical properties arise from these two features. The polysaccharide backbone can be considered as a condensation polymer of monosaccharides. The resulting glycosidic linkage (α or β) will partly determine the intrinsic stiffness of the polymer since the monosaccharidic units largely exist in the ring form (pyranose, etc.). Therefore, the only rotational possibility of the backbone is around the glycosidic bond. Similarly, the spatial distribution of the anionic groups and their degree of ionization will contribute to the conformation of the molecules due to the mutual electrostatic repulsion of the negative charges. In addition, both the conformation of a single molecule and the secondary structure of acidic polysaccharides in concentrated solutions, gels, or in the solid state will be influenced by the large amounts of inter- and intra-molecular hydrogen bonding among the abundant polar sites; e.g., —OH, —C—O—C, —C=O, NH₂, —COO⁻, —SO₃⁻.

The acidic polysaccharides occur mostly as extracellular material. Their main physiological role is to bind together cells and organs. In this sense their function is structural, providing toughness and flexibility in the connective tissues of animals, strength to the cell wall of plants and microorganisms, and largely acting as the cementing material of ground substance between cells. Since all substances going from cell to cell must pass through such ground substances, the chemical composition and physical state of the acidic polysaccharides may well influence the metabolism of cells. Physiological and pathological processes, such as control of electrolytes and water in extracellular fluids, lubrication, calcification, and wound healing, have been associated with the acidic polysaccharides matrix.

Many of the acidic polysaccharides occur in nature as protein and lipid complexes (especially those from animal tissues). In the present treatment, attention is focused on the carbohydrate moieties only. The nature of these complexes is not well known, although recently a large amount of structural work has been done on the carbohydrate–protein linkage. It seems that most of these are glycosidic bonds between the hydroxyl groups of the sugar moieties and serine or threoninine residues on the proteins (*1–5*). Ester linkages (*6,7*) have also been demonstrated. The main reason, however, for the restrictive treatment of this review is to provide a unified picture on the role and physical-chemical properties of acidic polysaccharides in animal and plant tissues. Hence, only compounds will be treated here in which the polymeric nature of the carbohydrate moiety is well established (i.e., they occur not as short oligomeric side chains of a protein core) and the main physical-chemical properties can be interpreted as largely the contribution of the acidic polysaccharide chains.

II. CHEMISTRY (STRUCTURE AND ORIGIN)

The most important acidic polysaccharides, the physical chemistry of which will be dealt in the subsequent paragraphs, can be divided into three groups according to their origin: (1) animals, (2) plant, and (3) derivative acidic polysaccharide.

A. Animal

Acidic polysaccharides from animal tissues belong to the mucopolysaccharides. This term was introduced by Meyer (*8*) to describe "hexosamine containing polysaccharides of animal origin." Obviously not all the mucopolysaccharides are acidic polysaccharides, but most of the compounds described below are treated under such group names in standard textbooks and monographs on the subject (*9–12*).

Hyaluronic acid is one of the most important polyelectrolytes of the connective tissues. It was first isolated by Meyer and Palmer (*13*) from vitreous humor of cattle eyes. It is widely distributed in different connective tissues, synovial and vitreous fluids of animals, and in some bacteria (*10,12,14,15*). Meyer has demonstrated (*16*) that it consists of equimolar quantities of D-glucuronic acid and 2-acetamido 2-deoxy-D-glucose. Subsequent works, mainly by Meyer and his group (*17*) and Jeanloz and coworkers (*18*), have established the primary structure (Fig. 1). The repeating unit of the chain polymer is a disaccharide, hyalobiuronic acid, in which the C-3 of the glucosamine moiety is connected by β-glycosidic linkages to the C-1 position of the glucuronic acid [2-acet-

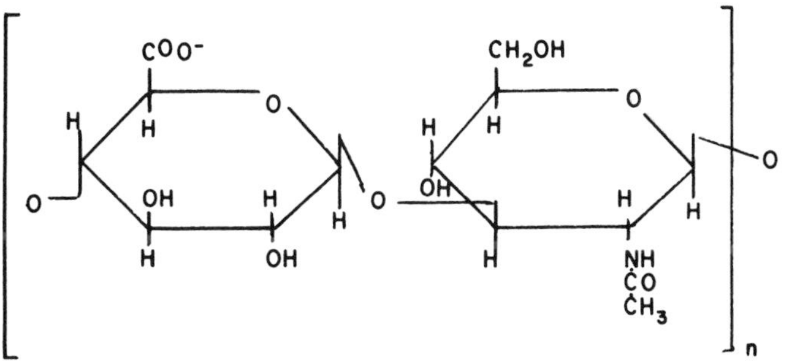

Fig. 1.   Hyaluronic acid.

amido-2-deoxy-3-O-(β-D-glucopyranosyl uronic acid)-D-glucose]. The polymer is made up of the 1-4-β-glycosidic linkages of the hyalobiuronic acid repeating group (12). Chondroitin is similar to the hyaluronic acid in the primary structure except that the hexosamine moiety is galactosamine instead of glucosamine. It was isolated by Meyer et al. (19) from bovine cornea, and on the basis of acidic and enzymic hydrolysis products it proved to be a polymer of $(1 \rightarrow 4)$-O-β-D-glucopyranosyl-uronic acid-$(1 \rightarrow 3)$-2-acetamido-2-deoxy-β-D-galactopyranose.

More frequent acidic polysaccharides of the connective tissues are chondroitin-4-sulfate and chondroitin-6-sulfate which are also called chondroitin sulfate A and C, respectively. Chondroitin is thought to be the biological precursor of chondroitin sulfates A and C. Chondroitin-4-sulfate (A) was first isolated from cartilage (20). Structural studies by Levene (21), Wolfrom (22), and Meyer (23) proved that the repeating unit is a disaccharide chondrosin in which the glucuronic acid is bound in a 1-3-β-linkage to the galactosamine moiety and the galactosamine is sulfated at the C-4 position (Fig. 2).

A partially sulfated chondroitin-4-sulfate was isolated by Bettelheim and Philpott (24) from bovine trachea. It was proposed that this highly crystalline chondroitin sulfate fraction was stereospecifically sulfated on an every third chondrosin moiety (25).

Chondroitin-6-sulfate (C) was first isolated by Meyer and Palmer (26) from umbilical cord. The only difference between chondroitin-4-sulfate and -6-sulfate is the position of the sulfate on the galactosamine moiety; in the first case it is in the C-4 position and in the second case it is in the C-6 position (27–29).

A chondroitin-6-sulfate preparation, which contained higher sulfate concentrations than chondroitin sulfate C, was isolated from shark cartilage (30,31). It was shown that this chondroitin-6-sulfate contained addi-

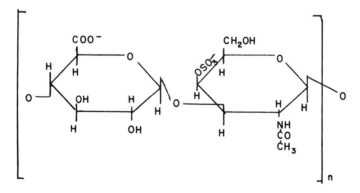

Fig. 2.   Chondroitin-4-sulfate.

tional ester sulfates on the uronic acid moiety (*32,33*). This shark cartilage chondroitin sulfate is also called chondroitin sulfate D (*12*).

Another acid polysaccharide of the connective tissue is called dermatan sulfate, but in the earlier literature it may appear under the names chondroitin sulfate B or β-heparin. Since it is different both chemically and physiologically from chondroitin sulfates A and C, we follow here the nomenclature adopted by Jeanloz (*11*). It was first isolated from pigskin by Meyer and Chaffee (*34*). It differs from the chondroitin sulfates in that the uronic acid moiety is not a D-glucuronic acid but L-iduronic acid (*35–37*). The repeating unit of the dermatan sulfate is therefore (1 → 4)-*O*-α-L-idopyranosyluronic acid-(1 → 3)-2-acetamido-2-deoxy-4-*O*-sulfo-β-D-galactopyranose (Fig. 3).

However, it was found that dermatan sulfate preparations always contain some D-glucuronic acid besides the L-iduronic acid. At first this was believed to be contamination of the sample by chondroitin-4-sulfate. Later it was proven by Fransson and Rodén (*38,39*) that both uronic acids are integral parts of the same chain, and the partial degradation of this acidic polysaccharide by testicular hyaluronidase yields fragments with D-glucuronic acid in the newly formed nonreducing termini. Upon fractionation of degradation products a tetrasaccharide was isolated which contained the sequence: glucuronic acid-*N*-acetygalactosamine-4-sulfate-iduronic acid-*N*-acetylgalactosamine-4-sulfate. This showed that at least in certain parts of the acidic polysaccharide chain the glucuronic acid and iduronic acid may appear in alternating sequence. The segment where this occurs is probably close to the dermatan sulfate–protein linkage (*40*) since analysis of this region indicated that only glucuronic acid appears in the polysaccharide sequence adjacent to the protein: glucuronic acid-*N*-acetylgalactosamine-4-sulfate-glucuronic acid-galactose-galactose-xylose-serine-peptide.

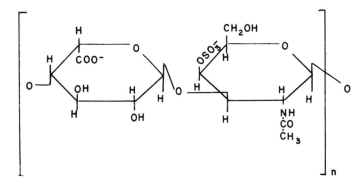

Fig. 3.   Dermatan sulfate.

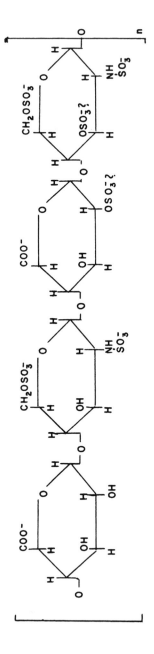

Fig. 4. Heparin.

A highly sulfated acidic polysaccharide, heparin, was isolated in 1916 by McLean from liver (*41*). It had very high anticoagulant activity and it occurs in liver muscles, lung, heart, kidney, spleen, and blood. In spite of the large amount of structural investigation performed on heparin, its primary structure is still not fully known (Fig. 4). The repeating unit is composed of equimolar quantities of D-glucosamine and D-glucuronic acid residues. Its sulfate content varies from 5–6.5 sulfate residues per tetrasaccharide (*11,42,43*). One of the sulfate groups is on the glucosamine-*N*-sulfate moiety (*44–45*). The other sulfate groups are most probably located as *O*-esters. The glucosamine moiety is considered fully *O*-sulfated at the C-6 position (*46*). On the basis of the behavior of heparin towards alkali hydrolysis, it was suggested that the remaining *O*-sulfate groups may be located at the C-2 position of the uronic acid residue and at the C-3 position of the hexosamine moiety. However, the degree of sulfation on these two positions varied from sample to sample, and that may cause the variability of the total sulfate content reported in the literature (*42*). The glycosidic linkage in heparin is mainly the (1 → 4)-α-glycosidic linkage (*47,48*). However, there is still a possibility that some glycosidic linkage at the 1 → 6 position provides branched structure, and some β configuration cannot be completely excluded. Another uncertainty is whether in heparin there are other uronic acids besides the D-glucuronic acid.

Although heparin specimens isolated from various terrestrial mammals are substantially identical, the heparin isolated from whale organs warranted a separate name, ω-heparin. Nearly one fourth of the hexosamine in the ω-heparin is *N*-acetyl-D-glucosamine. Yosizawa (*49*) demonstrated on the basis of biological activity and molecular weights, that the glucosamine content is not the result of a simple contamination by heparitin sulfate. Further differences between heparin, ω-heparin, and heparitin sulfate were demonstrated by electrophoretic, NMR, and ORD measurements (*50*).

Jorpes and Gardell (*51*) have isolated a dextrorotatory mucopolysaccharide from beef liver after the removal of heparin. This mucopolysaccharide, which is called heparitin sulfate (Fig. 5), contained equimolar amounts of glucosamine, uronic acid, acetyl, and sulfate residues. On the basis of its composition, it was suggested that the repeating unit may be a tetrasaccharide which contains one *N*-acetyl, one *N*-sulfate, and one *O*-sulfate group (*52*). Heparitin sulfate may be a block copolymer of disaccharides in which the glucosamine moiety carries *N*- and *O*-sulfates and of disaccharides having *N*-acetyl glucosamine only (*53*). It was reported that the heparitin sulfate may contain (1 → 6)-uronysyl-hexosamine linkage besides the (1 → 4) reported earlier (*54*).

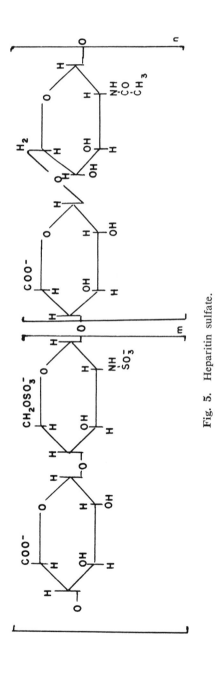

Fig. 5. Heparitin sulfate.

The idea that heparatin sulfate is a block copolymer rather than a co-polymer of sulfated and acetylated hexosamine-uronic acid disaccharides has been reinforced when Knecht et al. (55) found that, especially around the region where the acidic polysaccharide is bound to the protein, it consists largely of N-acetyl glucosamine-uronic acid repeating units. Furthermore, two fractions of heparitin sulfates were isolated from patients with Hurler's syndrome. One resembled the heparatin sulfates isolated from aorta while the second fraction resembled heparin, being highly sulfated. In Hurler's syndrome, heparitin sulfate as well as dermatan sulfate accumulate in the cells, presumably because they cannot be metabolized. Therefore, the heterogeniety of the heparitin sulfates in Hurler's syndrome may indicate that they originate from the partial degradation of higher molecular weight block copolymers of the normal cell constituents. The heterogeneous nature of heparitin sulfate has been shown also by the widely differing N-acetyl glucosamine content, depending on the source (56).

Keratan sulfate also called keratosulfate, Fig. 6, was first reported in bovine cornea (57) where it constitutes about half of the total muco-polysaccharide content. The repeating unit was shown to be N-acetyl-lactosamine which is sulfated in C-6 of the glucosamine moiety. It is polymerized through $1 \rightarrow 3$-β-glycosidic linkages (58). The corneal keratansulfate, also called keratan sulfate I, differs from skeletal keratan sulfate (II) in its relative stability toward alkali. It was proposed that the carbohydrate–peptide linkage in keratan sulfate II is via an O-glycosidic bond to threonine and serine while in keratan sulfate I the bond is either a glycosylamine linkage between carbohydrate and asparagine and glutamine or an amide linkage between the glucosamine and aspartic and glutamic acids. Keratan sulfate I and II differ slightly in their minor constituents; keratan sulfate I having smaller sialic acid and methylpentose content than II (59). Some keratan sulfate preparations from senile human cartilage had a greater than $1:1$ sulfate to hexosamine ratio. In these cases the galactose moiety is also sulfated at the C-6 position (60).

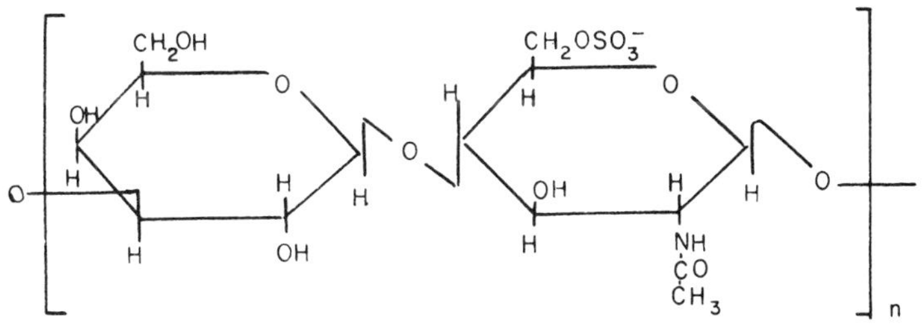

Fig. 6. Keratan sulfate.

The sulfate groups appear to be randomly distributed between C-6 positions of D-galactose and N-acetyl-D-glucosamine, with the larger share of sulfate groups on the glucosamine. Keratan sulfate II from human rib cartilage is also more branched than keratan sulfate I, with sialic acid, fucose, and some galactose being present in the branches individually or in small groups (61).

Keratan sulfate isolated from different human and bovine tissues showed a great variation in the sulfate:hexosamine molar ratio. This varied from a low 0.46 in bovine nasal septum cartilage to a high of 1.60 in human nucleus pulposus. Furthermore, keratan sulfate even from a single tissue showed heterogeneity in the degree of sulfatation (62).

A number of sulfated polyglucans have been isolated from lower animals. From the odontophore of the marine snail, *Busycon canaliculatum,* a cellulose sulfate was isolated (63). From the hypobronchial gland of the same animal an acidic polysaccharide containing D-glucosamine, D-galactosamine, and ester sulfate was obtained (64). A (1-4)-β-linked glucan sulfate was isolated from *Charonia lampas* (65). The hypobronchial mucin of the whelk *Buccinum undatum* L. contains an acidic polysaccharide loosely bound to a glycoprotein through salt linkages (66). The acidic polysaccharide itself is linked to a peptide moiety, but whether this is an ether linkage to serine or ester linkage to aspartic acid has not been established conclusively (67). The acidic polysaccharide itself is a β-(1-4)-linked polyglucan, probably sulfated on the C-2 or C-3 position (68).

B. Plant

Among the acidic polysaccharides occurring in plants, the pectic substances are the most common. They usually occur in the cell walls of higher plants and the intercellular cementing material (middle lamellae) and in the cells sap. A wide variety of materials is hidden under the heading "pectic substances." The common feature in each is the polygalacturonic acid chain. In its purest form, the pectic acid (Fig. 7), is composed of α-D-galactopyranosyl uronic acid in the 1-4-α-glycosidic linkage (69). The naturally occurring compound or pectin is a partly or fully esterified pectic acid in which galactose and arabinose units also occur. These neutral sugar moieties are believed to be copolymers, probably linked to the galacturonic acid moieties by covalent linkages. However, pure polygalacturonic acid is not a degradation product, an artifact of isolation, since it has been isolated from plant extracts without any degradative processes by simply passing the extracts through a cation exchange column, and by ultracentrifugation (70). Similarly, polygalac-

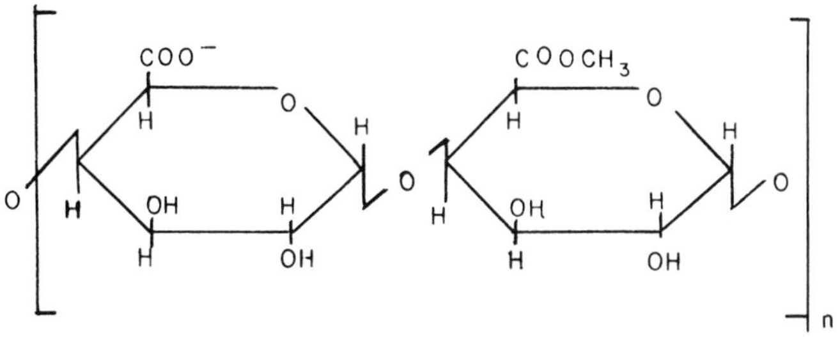

Fig. 7. Pectinic acid.

turonic acid was also isolated by gel filtration (71) which also yielded several fractions; one contained galacturonic acid and arabinose, the other contained only neutral sugars of galactose and arabinose. Other sugar moieties, such as sorbose ramulose and xylose, have been reported. Under the name pectinic acid, we usually mean polygalacturonic chains in which the methoxyl groups have been removed by acid, alkaline, or enzymic treatment to values below 50%.

Another important group of acidic polysaccharides of plant origin, and in which uronic acid acts as a repeating unit, are the alginates. These are isolated mainly from brown algae where they perform functions similar to pectins in higher plants. The repeating unit in alginic acid has been shown to be D-mannuronic acid (72), and it is accepted that the polymer is composed of anhydro-β-D-mannuronic acid in $1 \rightarrow 4$-glycosidic linkages (73).

Some alginic acids have more complex structures as was proved by Hirst and Rees (74). They isolated and methylated a polyuronide from *Laminaria cloustoni,* the hydrolysis product of which yielded 2,3-di-*O*-methyl-D-mannose and 2,3-di-*O*-methyl-L-gulose. This suggests that the polyuronide consisted of 1-4-β-linked mannuronic acid and 1-4-α-linked guluronic acid as the major constituents.

Haug et al. (75) studied the uronic acid sequences in commercial alginic acid from *Laminaria digitata.* They report, on the basis of the electrophoretic behavior of degradation products, that alginic acid is probably a block copolymer of polymannuronic acid (M-blocks) and polyguluronic acid (G-blocks). In between these blocks are fragments composed of alternating M-G-M-G residues.

An even more complex acidic polysaccharide was isolated from the seaweed *Cladophora rupestris* (76). The anionic groups on this polymer were L-arabinose-3-sulfate and D-galactose-6-sulfate in addition to xylose. The molar ratio of xylose:galactose:arabinose were 1:3:3. The polymeric linkages of galactose moieties were 1-3 and 1-6, those of arabinose 1-4 and 1-5, and of xylose 1-4. Extract of brown seaweed, *Ascophyllum*

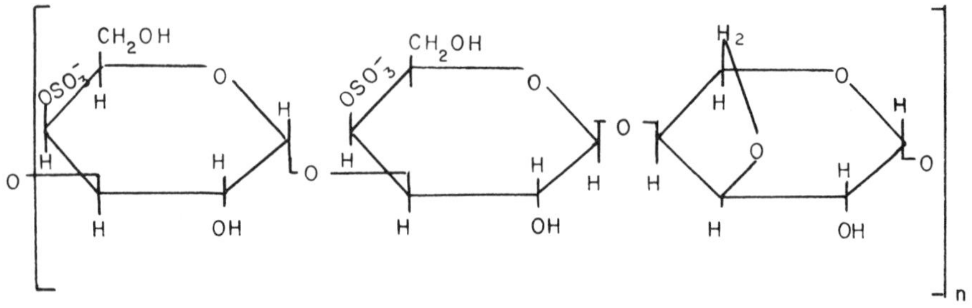

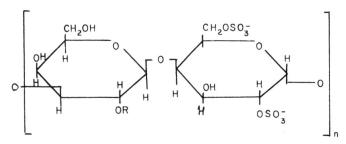

Fig. 8.   κ-Carrageenan.

nodosum, contains a sulfated polyuronide, ascophyllan, in which the backbone is a polyglucoronic acid chain with sulfated mono- and disaccharide side chains (*77*).   The disaccharidic side chain proved to be a 3-*O*-β-D-xylopyranosyl-L-fucose moiety (*78*).

Carrageenans are extracted from lichens such as *Chondrus crispus, Ch. ocellatus,* and *Gigartina stellata.*   In the form of potassium salt they separate into κ and λ forms.   The κ form is a polymer of galactose-4-sulfate and 3,6-anhydro-galactose units (*79*) of the formula given in Fig. 8.   The λ form is more highly sulfated and the proposed repeating unit is shown in Fig. 9 (*80*).

Lately, however, Pernas et al. have shown that the carrageenans are not composed of only the two chemically well-defined fractions κ and λ, but the term carrageenan has to encompass a series of molecules of different chemical composition and of different solubility (*81*).   Therefore, the κ and λ designations should apply only to the particular fractionation process originally devised (*82*).

Some of the acidic polysaccharides from plant tissues come under the somewhat vague terminology of hemicelluloses.   Under this group name, we include polysaccharides of somewhat lower molecular weight which occur in plant tissues together with cellulose mainly in the secondary wall of cells.   They can be isolated from the original or from the delignified

Fig. 9.   λ-Carrageenan.

material by aqueous or aqueous alkali extractions. The acidic character of hemicelluloses is less than in the previous polyuronides. This is due to the relatively low uronic acid content of these hemicelluloses. Hard woods contain O-acetyl-(4-O-methylglucorono-) xylans (83). The complete structure of xylan isolated from beechwood has been carried out by Aspinall et al. (84). The (glucorono-) xylan, according to this work, is composed of a linear polymer of (1 → 4)-linked β-xylopyranose residues, and about every tenth of this residue has a branched side chain with a terminal 4-O-methyl-α-D-glucuronic acid residue attached to the C-2 position of the xylose. Many other (4-O-methylglycorono-) xylans have been isolated from angiosperms (83) in which the number of glucuronic acids side chains varied from 1 per 5 to 1 per 15 xylose residues. In most other respects the xylose repeating units in the backbone and in the side chains were the same. Some of the xylan backbones seemed to be branched.

Acidic polysaccharides from hemicelluloses of soft wood present a somewhat more complicated picture (83). One such soft wood xylan was isolated by Dutton and Smith (85) from western hemlock. The fundamental repeating units of this polysaccharide consist of (1 → 4)-linked β-D-xylopyranose residues in which two types of side chains occur. One of the side chains is 4-O-methyl-α-D-glucuronic acid and this is linked in a 1 → 2 linkage to the polymer backbone. The other side chain is a (1 → 3)-linked L-arabinouranose residue. About 13 xylose units in the backbone have two of these side chains. Similar arabino-(4-O-methyl glucorono) xylans have been isolated from gymnosperms in which the xylose residue per acid side chains varied from 4.3–9 (83).

Many plant species produce water-soluble gums which contain heteropolysaccharides of complex structure and in most of these gums, uronic acid moieties, are integral parts of the structure. Mostly these are glucuronic acid moieties, but in some compounds digalacturonic acid has been found (86). Most of these gums are excreted by trees upon physical injury. Gum arabic and gum tragacanth are the most important from industrial points-of-view. Gum arabic contains L-arabinose, L-rhamnose, D-galactose, and D-glucuronic acid in approximately 3:1:3:1 ratios. The proposed structure consists of a backbone in which the galactose units are polymerized in 1-3 linkages and every galactose carries a side chain at the C-6 position. They are two types of side chains in equimolar quantities; one consists of a trisaccharide in which the D-galactose unit is the terminal group, and it is linked in a 1 → 3 to an L-arabinose which in turn is linked, also in a 1-3 linkage, to another L-arabinose unit. This trisaccharide side chain is linked in a 1-6 linkage to the galactan backbone. The other side chain contains a branched tetrasaccharide unit in which a terminal L-

arabinose group is linked in a 1-4 linkage to a D-galactose moiety.   This
D-galactose moiety is linked in a $1 \rightarrow 6$ linkage to a main backbone, and
in its 3 position it contains an L-rhamnose moiety.

More recent experiments by Anderson and his coworkers have shown
individual variations between the gums extruded by different species of
acacia trees.

Thus, *Acacia arabica* gum has a very low rhamnose content (0.4%)
and contains a highly branched galactan backbone.   Many of the branch-
ing points start with galactose unit linked in $1 \rightarrow 6$ β linkages to the main
backbone in which $1 \rightarrow 3$ and $1 \rightarrow 6$ β-linkages vary randomly.   Some of
the side chains end in D-glucuronic acid, which is linked by either $1 \rightarrow 4$
α- or $1 \rightarrow 6$ β-glycosidic linkage to the galactose branching points.
Others end in 4-*O*-methyl-D-glucuronic acid, similarly linked with 1-4 α-
and 1-6 β-linkages to the galactose branching points.   Some 4-*O*-methyl-
D-glucuronic acid moieties are linked directly to the backbone.   Similarly,
the longer arabinose-containing side chains are linked either to galactose
branching points or directly to the backbone.

The latter feature seems to be a unique property of *A. arabica* gum,
in contrast to other species (*87*).   For example, *A. senegal* gum has all
its arabinose side chains linked to galactose branching point (*88,89*).
The arabinose side chains of *A. acacia* gum are as long as 6 monosaccharidic units while those of *A. senegal* are shorter.   The arabinose side
chains contain various linkages such as 1-3- and 1-2-β-L-arabinopyranosyl
and 1-3- and 1-2-β-L-arabinofuranosyl linkages.

Further species of acacia gums showed that the gum of *A. lacta* is similar to that of *A. senegal,* but it has a greater proportion of 1-3-β linkage
in the backbone (*90*).   The gum of *A. dreponolobium* is not completely
water soluble, and three fractions with different primary composition have
been isolated: (1) a cold-water soluble; (2) a salt soluble; and (3) an
alkali soluble fraction.   The cold-water soluble fraction, which makes up
80% of the total gum, is very similar to the gum of *A. acacia* (*91,92*).

Similar structural studies on lemon gum indicated that it consists of
36% D-galactose, 33% L-arabinose, 5% L-rhamnose, 7% D-glucuronic
acid, and 16% 4-*O*-methyl-D-glucuronic acid.   The backbone of the
molecule is similar to those found in Acacia gum; namely, it consists of a
branched galactan framework in which $1 \rightarrow 3$ and $1 \rightarrow 6$ β linkages vary
randomly.   The following side chains were found (starting from the
branching point and ending with the terminal sugar moiety): $6 \leftarrow 1$
β-L-Arabinopyranosyl $4 \leftarrow 1$ 4-*O*-methyl-D-glucuronic acid; $6 \leftarrow 1$ β-D-
galactopyranosyl $4 \leftarrow 1$ α-D-glucuronic acid; $3 \leftarrow 1$ β-D-galactopyanosyl
$3 \leftarrow 1$ β-D-galactopyranosyl $4 \leftarrow 1$ 4-*O*-methyl-α-D-glucuronic acid; $3 \leftarrow 1$
β-L-arabinofuranosyl (2 and) $3 \leftarrow 1$-β-L-arabinofuranosyl $3 \leftarrow 1$-L-arabino-

pyranosyl 4 ← 1 L-rhamnose; 3 ← 1-L-arabinofuranosyl 3 ← 1-L-arabino-furanosyl 3 ← 1-L-arabinopyranosyl 4 ← 1 4-O-methyl-D-glucuronic acid. The last two tetrasaccharidic side chains are bound to the backbone not only by the 1-3-β linkages given above, but also by 1-6-β and 1-4-β linkages (*93*).

Acacia gum and black wattle gum have the same components but the L-arabinose, L-rhamnose, D-galactose, and D-glucuronic acid are in 6:1:5:1 proportions. Gum tragacanth is similar to gum arabic but it is more complex, containing L-arabinose, D-xylose, L-fucose, D-galactose, and D-galacturonic acid moieties. Its detailed structure is not known.

A nonsulfated acidic polysaccharide resembling chondroitin has been isolated from *Bacillus subtilis* (*94*). It contains equimolar quantities of N-acetylgalactosamine and glucuronic acid. On the basis of positive optical rotation, resistance to testicular hyaluronidase, and ease of acid hydrolysis, it was proposed that the repeating unit is a 3-O-α-glucuronosyl-N-acetylgalactosamine moiety. The galactosaminidic linkage is still unknown.

Another acidic polysaccharide isolated from bacteria is colominic acid. It was found in *E. coli* K_{235} culture, and upon acid hydrolysis it yielded crystalline N-acetylneuraminic acid (*95,96*). It is believed that the polysaccharide is mainly a polymer of N-acetylneuraminic acid although other constituents may be also present. Neither the degree of polymerization nor the polymeric linkage is known at present.

Some bacteria, such as *E. coli, S. typhosa,* and *P. bakerup* have a polymer which is commonly called Vi antigen. The repeating unit is 2-N-acetamide-2-deoxy-D-galacturonic acid (*97*), and it is believed that the monosaccharides are joined in an α-(1-3) glycosidic linkage (*98*). Pneumococci have acidic polysaccharides and the structure of some have been partly elucidated. Type III contains a polymer of 4-O-β-D glucuronosyl-D-glucose moieties which are joined by (1 → 3) linkages (*99*). Type V has a repeating unit of 3-O-β-D-glucuronosyl-N-acetyl-L fucosamine, but it also contains some D-glucose units (*100*). Type VII has a polymer in which the repeating unit is D-glucuronic acid-[β-(1-4)-D-glucose]$_2$-α-(1-4)-D galactose (*101*). A polyuronide resembling alginic acid was isolated from Pseudomonas. It contained both mannuronic and guluronic acids. The polymer is esterified about 25–50% but the ester groups are O-acetyl groups (*102*).

Carlson and Matthews (*103*) proposed that the polyuronides isolated from different mucoid types of *Pseudomonas aeruginosa* are in essence heterogeneous polymer chains. They were able to isolate pure polymannuronic acid. They suggested that the wide variation in the guluronic acid content of the polyuronide isolated from different types of *P. aeru-*

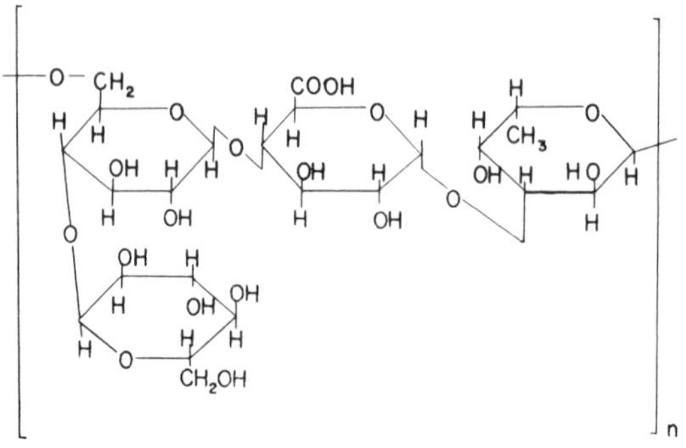

Fig. 10.   Repeating sequence of the acidic polysaccharide of *A. aerogenes*.

ginosa is the result of different mixtures of polymannuronic and poly-guluronic acids.

An extracellular acidic polysaccharide was isolated from *Arthrobacter viscosus* by Jeanes et al. (*104*).   The polysaccharide contains D-glucose, D-galactose, and D-mannuronic acid in 1:1:1 molar ratios.   Siddiqui (*105*) has shown that this is a straight chain polymer of the trisaccharide *O*-(β-D-mannopyranosyluronic acid)-(1 → 4)-*O*-β-D-glucopyranosyl-(1 → 4)-D-galactose.   The trisaccharides themselves may be linked through 1 → 4-β linkages.

The acidic slime polysaccharide isolated by Aspinall et al. (*106*) was first thought to be a very complex structure made of D-glucose, D-glucuronic acid and L-fucose moieties in a nonrepeating sequence.   Lately Conrad et al. (*107,108*) have shown that the repeating unit is a tetra-saccharide (Fig. 10).

C. Derivative Acidic Polysaccharides

Derivative acidic polysaccharides are usually made from neutral poly-saccharides, such as cellulose and amylose, by the action of certain chem-ical agents.   One of the most important derivative acidic polysaccharide productions is the manufacture of carboxymethyl cellulose.   Although numerous industrial patents have been taken out on its manufacturing process, essentially it is made by treating cellulose in caustic soda to pro-duce steeping cellulose which is then etherified with $CH_2ClCOONa$ (*109–111*).   Its degree of substitution varies with the industrial use.   A degree of substitution of 1.0 means one carboxymethyl group per mono-mer.   The C-6 position is preferentially substituted.   For laundry use, the degree of substitution is usually 0.6–0.8.   Other uses are found in drilling muds, pharmaceuticals, and in the textile and paper industries (*112,113*).

Other cellulose ethers belonging to the polyelectrolytes are carboxyethyl cellulose and polysulfated carboxymethyl cellulose. The latter and the cellulose sulfates have received attention as possible anticoagulants due to their structural resemblance to heparin (114).

Other ethers and esters with polysaccharidic backbones include dextran sulfates, starch sulfates, and phosphates and ethers in which both the degree of anionic substitution and the degree of branching of the polysaccharidic backbone can be varied (115,116).

The action of periodic acid upon cellulose, followed by treatment with chlorous acid, produces an oxycellulose in which the secondary hydroxyl groups have been converted into carboxyl groups with the simultaneous rupture of the pyranose ring. This is called the periodate-chlorite oxy-cellulose (117,118). An oxycellulose in which the primary alcohol groups of the cellulose are oxidized to carboxyl groups is produced by the action of nitrogen dioxide (119). Another type of derivative acidic polysaccharide is the different anionic forms of synthetic polyglucose prepared by Mora and his coworkers (120). Polyglucose sulfates with different degree of molecular weights and substitutions have been prepared (121). They are strong anticoagulant and showed different enzyme inhibitory potencies (122). Sodium carboxyl derivatives of polyglucose have also been prepared (123) and they also showed different biological activities.

A completely different group of derivative acidic polysaccharides are the graft copolymers in which synthetic organic polymers are grafted onto a naturally occurring acidic polysaccharide backbone. For example, a low molecular weight "living" polystyrene which is an anionic polymeric species has been grafted onto a xylan backbone. The xylan was methylated first in order to eliminate the chemical activity of the OH groups. The methyl ester group of the glucuronic acid then coupled with two polystyrene chains. Such reaction products can be prepared with any methylated polyuronide (124).

III. MOLECULAR PARAMETERS OF SINGLE MOLECULES

The molecular parameters of a single polyelectrolyte can be obtained theoretically by different statistical-mechanical treatments of certain models. Experimentally one approaches the problem through the dilute solutions of the polymers. Polymer solutions are far from ideal, but at high dilutions they approach a quasi-ideal situation and the results usually can be extrapolated to infinitely dilute solution. In this process one assumes that the interaction between polymer molecules will be diminished and at infinitely dilute solutions only a single molecule will influence the behavior of the system. However, in the case of polyelectrolytes this simple assumption is complicated by the fact that upon dilution the degree of dissociation of the counterions is changing and, therefore, the physical-chemical properties of the solutions are affected also by the number of counterions independently present.

A. Molecular Weights and Molecular Weight Distributions

The most important molecular parameter that can characterize a given polymer is the molecular weight. If all the molecules have the same molecular weight (monodisperse system), the molecular weights determined by different types of measurements will give the same value. However, in the case of polydisperse systems, different techniques will give different values. The simplest physical-chemical properties used for molecular weight determinations are those of the colligative properties. Vapor pressure lowering, boiling point elevation, freezing point lowering, and osmotic pressure measurements are among the colligative properties most frequently used. The first three techniques are not very suitable for molecular weight determinations of polymers because, due to the large molecular weight and low solubility, the number of particles present in a unit volume is extremely low. Osmotic pressure, on the other hand, is sufficiently sensitive to yield good results even at low concentrations, and it is therefore frequently used in determining molecular weights. However, lately vapor pressure lowering measurements have also been used since commercially available instruments appeared on the market under the name of "vapor phase osmometers." Osmotic pressure, or any colligative property, is concentration-dependent and at a constant temperature can be expressed by the Flory equation (125)

$$\pi/c = (\pi/c)_0(1 + A_2c + GA_2^2c^2 + \cdot \cdot \cdot) \tag{1}$$

π is the colligative property measured, the concentration c is in grams/liter, $(\pi/c)_0$ is the limiting value at zero concentration, and A_2 and GA_2^2 are virial coefficients characteristic for the solute–solvent and solute–solute interactions. The molecular weight is obtained from the limiting value

$$(\pi/c)_0 = \frac{RT}{M} \tag{2}$$

where R is the gas constant, T is the absolute temperature, and M is the average molecular weight of the polymer. In the case of polyelectrolytes the complicated factor is that the number of particles will include counterions and, therefore, the molecular weight will be an average of the polyelectrolytes and the counterions. Previously, to avoid these difficulties, the polyelectrolytes were dissolved in an excess of neutral salt solutions and the osmotic pressures of these solutions were measured against a solvent which also contained the same concentration of salt.   The assumption of this procedure was that the presence of salt minimized the dissociation of the counterions.   It was recently found that polyelectrolyte solutions containing low molecular weight salts yield osmotic pressures (against a solvent) in which a simple additivity law is followed.   According to this empirical law, the sum of the effective ionic concentrations contributed independently by the polyelectrolyte and the salt is expressed by the total activity of ions.   Theoretical explanations for this simple additivity law have been advanced by different authors (126–130).

Light scattering has also been used to determine the molecular weights of polymers.   The intensity of light scattered at different angles in a polymer solution depends on the concentration of the solution, on the molecular weight and size of the polymer, and on the shape of the molecules.   In order to evaluate molecular weights, one again has to extrapolate to infinitely dilute solution, and this is done by the so-called Zimm plot (Fig. 11). C is the concentration of the polymer solution in grams/ milliliter, R_θ values are the corrected Raleigh ratios for unpolarized light

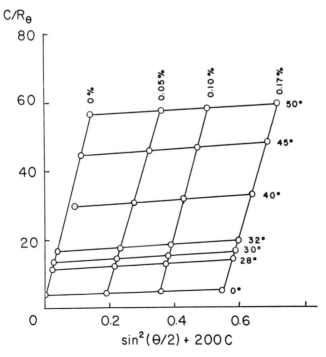

Fig. 11.   Zimm plot of carboxymethyl cellulose.

obtained from the scattering intensities at different θ angles, and k is used as an arbitrary constant to provide a suitable gridlike plot. The experimental points are extrapolated to both zero concentration and zero angle. These two lines intercept the ordinate at one point. When the double intercept of the Zimm plot is multiplied by an optical constant, K,

$$K = \frac{2\pi^2 n^2 (dn/dc)^2}{N\lambda^4} \qquad (3)$$

the reciprocal of the weight-average molecular weight is obtained. The parameters included in the K are the refractive index of the solvent n, the refractive index increment dn/dc, Avogadro's number N, and the wavelength of light λ (131).

Ultracentrifugal techniques are also used to obtain the molecular weights by applying the Svedberg equation (132)

$$M = RTS_0/(1 - \bar{V}\rho)D_0 \qquad (4)$$

where M is the molecular weight, R is the gas constant, T is the absolute temperature, S_0 is the extrapolated sedimentation coefficient, and D_0 is the extrapolated diffusion coefficient which has to be obtained independently. $\bar{V}$ is the partial specific volume and ρ is the density. The particular-average molecular weight obtained by the Svedberg equation depends on the selection of the sedimentation and the diffusion boundaries (133). The so-called Archibald technique of equilibrium centrifugation yields weight-average molecular weight which is related to the ratio of concentration C_2/C_1 at two points in the centrifuge cells at distances x_1 and x_2 from the axis of rotation.

In the latter case, however, it has been pointed out that most of the data in the literature may be incorrect because the calculations for molecular weights of polyelectrolytes in salt solutions used a dn/dc value which is too low, and therefore correction terms have to be employed (134,135). Furthermore, as Nichol et al. pointed out (136), the sedimentation equilibrium equations were developed for homogeneous solutes and are used with particularly good results in the case of proteins where deviation from homogeneity is minimal (137). In polydisperse systems, for example, the nonideality behavior of the solute introduces such effects as a maximum in the Schlieren diagram near the bottom of the cell. This maximum can be used to evaluate second and third virial coefficients. In some cases, e.g., hyaluronic acid, besides a sharp maximum a minimum also appears and tends to increase at the bottom indicating a rapidly sedimenting fraction. This must be the viscoelastic putty of hyaluronic acid isolated by Balazs in ultracentrifugation experiments (317). Sundelöf (138) proposed a convolution procedure which uses the integral transforms of sedimentation diffusion equilibrium data to obtain the poly-dispersity and different-average molecular weights.

These three techniques, together with viscosity measurements (see Section V.A), are usually used to obtain molecular weights and molecular weight distributions. The osmotic pressure and colligative property measurements give number-average molecular weights, the light-scattering and ultracentrifuge techniques give weight-average molecular weights, ultracentrifuge techniques can also yield Z-average molecular weights, while viscosity measurements give viscosity-average molecular weights. The relationship between these molecular weights are as follows:

$$M_n = \frac{W}{\sum\limits_{i=1}^{\infty} N_i} = \left(\sum_{i=1}^{\infty} M_i N_i\right) \Big/ \left(\sum_{i=1}^{\infty} N_i\right) \tag{5}$$

$$M_w = \left(\sum_{i=1}^{\infty} c_i M_i\right) \Big/ c = \left(\sum_{i=1}^{\infty} N_i M_i^2\right) \Big/ \left(\sum_{i=1}^{\infty} N_i M_i\right) \tag{6}$$

$$M_z = \left(\sum_{i=1}^{\infty} N_i M_i^3\right) \Big/ \left(\sum_{i=1}^{\infty} N_i M_i^2\right) \tag{7}$$

$$M_v = \left[\left(\sum_{i=1}^{\infty} N_i M_i^{(1+a)}\right) \Big/ \left(\sum_{i=1}^{\infty} N_i M_i\right)\right]^{1/a} \tag{8}$$

where M_i = molecular weight of species i, N_i = number of moles of species i, w = total weight of sample, and c_i = concentration. The parameter a in Eq. 8 is a shape factor (see Eq. 35). The molecular weight distribution of a sample can be obtained by fractionating the sample in mixed solvents or at different temperatures and determining the molecular weights of each of these fractions by any of the above-named techniques. Different nonuniformity coefficients representing poly-dispersity have been proposed (139,140).

Molecular parameters of single molecules of acidic polysaccharides are tabulated in Table I. This summation is not intended to be complete, but it does try to give a fair representation of work done in this field.

The first thing that is apparent in looking at the molecular weights obtained for different acidic polysaccharides by different techniques is the large degree of polydispersity which these polymers possess. This is obvious from data that gives the different number, weight, etc., average molecular weights of the same sample (148–150,168,169). It is also obvious from studies in which the original samples were fractionated and

TABLE I[a]

Physical Parameters of the Acidic Polysaccharides

Acidic polysaccharide	Source	Medium	Technique	M_n	M_w	M_z	A_2	dn/dc	R_G	L	Shape	Ref.
Hyaluronic acid	Vitreous body	$0.005\ M\ \text{PO}_4^{2-}$	U		8.6×10^4						Ellipsoid	141
		$0.1\ M$ NaCl	LS		$3.5\text{--}5 \times 10^5$				2000		Random coil	142
		$0.2\ M$ NaCl	LS		$7.7 \times 10^4\text{--}1.7 \times 10^6$			0.166 (546 mμ)	$540\text{--}1430$			133
		$0.2\ M$ NaCl	U		$1.0 \times 10^5\text{--}1.5 \times 10^6$							133
			LS + U		$3 \times 10^5\text{--}1.3 \times 10^6$					$250\text{--}480$	Rod random coil	148
			O	2.7×10^5								144
	Umbilical cord		U		$3\text{--}5 \times 10^5$							145
			LS + U		$5 \times 10^5\text{--}8 \times 10^6$					$480\text{--}700$	Rod random coil	148
	Synovial fluid		LS + U		$3 \times 10^5\text{--}10 \times 10^6$					$400\text{--}1000$	Random coil	143
	Ovarian cyst		O	$4.5 \times 10^5\text{--}1.2 \times 10^6$								146
Chondroitin-4-sulfate	Streptococcus	$0.1\ M\ \text{PO}_4^{3-}$	V		4×10^4							147
	Bovine nasal cartilage	$0.15\ M\ \text{PO}_4$ + $0.2\ M$ NaCl	O + LS	$1\text{--}4 \times 10^4$	$1.7\text{--}5 \times 10^4$			0.166 (546 mμ)			Random coil	148
Chondroitin-6-sulfate	Human chordoma	$0.15\ M\ \text{PO}_4$ + $0.2\ M$ NaCl	O + LS	$1.4\text{--}4.3 \times 10^4$	$1.8\text{--}15 \times 10^4$		2.0×10^{-3}				149
			O	3.5×10^4								148
Dermatan sulfate	Pig skin	$0.5\ M$ NaCl	U	2.3×10^4	2.7×10^4	4.1×10^4		0.160 (546 mμ)				160
	Beef lung	$0.15\ M\ \text{PO}_4$ + $0.2\ M$ NaCl	O + LS	$1\text{--}1.5 \times 10^4$	2.2×10^4							148
Keratan sulfate	Nucleus pulposus	$0.2\ M$ NaCl	U + LS		$1.5\text{--}2 \times 10^4$							151, 152
	Cornea		U + LS		$0.87\text{--}1.9 \times 10^4$							

Heparin	Hog lung		U	1.2×10^4							153
	Human lung		U	1.5×10^4							153
	Hog beef	0.1–1.0 N NaCl	LS + U	1.6×10^4				0.121 (436 mμ)			154,
		0.2 N NaCl	U + LS	$8–12 \times 10^3$				0.132 (546 mμ)			155 156
Heparitin sulfate	Hurlers Syndrome liver		O		$2.7–5.5 \times 10^3$						157
	Aorta				$24–29 \times 10^3$						55
Pectinic acid	Apple, plum, pear		U	$2.5–3.5 \times 10^4$							158
	Orange	0.2 N NaCl	U	$4–5 \times 10^4$				0.140 (436 mμ)		Rod	158
	Apple	0.2 N NaCl	U	5.8×10^4		8.9×10^4					159
	Citrus	0.155 M NaCl	O		$1.8–4 \times 10^4$		-0.4×10^{-3}		530–1650	Rod	159
Alginic acid	L. digitata		LS + U	6.9×10^4		11.9×10^4		0.115 (436 mμ)	1080	Stiff coil	159
		0.1 N NaCl	LS	8×10^5			-0.1×10^{-3}			Extended coil	161
Carrageenan			O + U	$1.1–5.3 \times 10^5$	$6–22 \times 10^4$					Rod	162
Agar-agar			LS	$8–40 \times 10^3$							163
Arabic acid	Gum arabic	0.02 N HCl	LS	1×10^6			3.2×10^{-3}	0.150 (540 mμ)	555	Stiff coil	164
	Gum arabic	1 N NaCl	LS	2.3×10^6					550	Compact coil	87
	Gum arabic	0.2 N NaCl	O		3×10^5						165
			U	4.1×10^5		5.2×10^5		0.145 (436 mμ)		Stiff coil	159
	Gum senegal	1 N NaCl	LS	5.8×10^5					739	Stiff coil	91
	Gum nubica	1 N NaCl	LS	8.7×10^5					340	Compact coil	91
Pneumococcal I		0.15 M NaCl	U	1.71×10^5							166
II			U	5.04×10^5							166
III			U	1.41×10^5							166
Teichoic acid	Staphylococcus Acid Extr.		U	2.3×10^4							167
			U	7.8×10^3							

TABLE I (Continued)

Acidic polysaccharide	Source	Medium	Technique	M_n	M_w	M_z	A_2	dn/dc	R_G	L	Shape	Ref.
Glucan sulfate	B. undatum	0.5 M NaCl			1.72×10^5							68
	C. lampas				1×10^6							65
Polyhexosamine sulfate	Snail		LS		2.5×10^6							64
Carboxy-methyl cellulose		0.2 M NaCl	O + U + V	2.9×10^5	1×10^6	1.83×10^6	13×10^{-4}	0.134 (433 mμ)	1075			168
		0.005 M NaCl					88×10^{-4}	0.138 (436 mμ)	2010			168
		0.005 M NaCl	O + LS	$1.2\text{-}2 \times 10^5$	$3.65\text{-}4.8 \times 10^5$		26×10^{-3}	0.154 (436 mμ)	3350			169
		0.5 M NaCl					1.3×10^{-3}		2370			169
		0.05 M NaCl							820			170
		0.2 M NaCl	O + U + LS + V		$1.5\text{-}10 \times 10^5$		1.0×10^{-3}	0.136 (546 mμ) 0.131 (546 mμ)	290			170
Amylose sulfate			O	3.7×10^5								171
Guaran sulfate			O	1.6×10^5								171
Dextran sulfate	Degree of sulfation 2.3		O	1×10^6								172
	2.0		O	3×10^5								
	1.4		O	6×10^4								
	Linear		LS + U	4×10^4								173
	Branched		LS + U	6×10^4								173
Carboxy-polyglucose			ENDG	1.3×10^4								123

a O = osmotic pressure; LS = light scattering; U = ultracentrifuge; V = viscosity; ENDG = end group determination; M_n, M_w, M_z refer to Eqs. 5–8; A_2 refers to Eq. 1; dn/dc refers to Eq. 3; R_G is the radius of gyration in Å; and L is the length of the rod in Å.

the molecular weight of each fraction determined individually (*133,156, 160*). In this respect the acidic mucopolysaccharides differ from proteins which are largely monodisperse.

A second feature apparent in this tabulation is that the molecular weight of an acidic polysaccharide depends on the source and on the technique of isolation. Hyaluronic acid is a good illustration of this point. While the vitreous body usually contains hyaluronic acid of lower molecular weights compared to those from umbilical cord or synovial fluids, it is also true that the different preparations contained different amounts of residual proteins varying from 3–15%. The importance of the technique of isolation is emphasized by the fact that hyaluronic acid preparations undergo an oxidative–reductive depolymerization in the presence of many compounds (*174–176*). There is also a question of depolymerization by enzymes during the preparations or *in situ* due to pathological conditions, e.g., synovial hyaluronic acid from certain types of arthritic conditions (*177–179*). Similar considerations apply to the other acidic polysaccharides.

B. Size and Shape of the Molecule

Besides the molecular weights and molecular weight distributions, the other important parameters describing the single molecule are the size and shape factors. The size can be obtained from light-scattering measurements (*131*) either as the radius of gyration, R_G, or the end-to-end distance of the molecule which, in the case of a rod shape, is the length of the molecule (L). Once the radius of gyration is obtained from the zero concentration line slope and the intercept of the Zimm plot, the shape of the molecule can be estimated by comparing the values of the extrapolated points on the zero concentration lines to theoretical function of the particle scattering factor $P(\theta)$ for standard shapes such as compact sphere, random coil, and rod. Shapes of the molecule can also be estimated from axial ratios (length over width) which are obtainable from ultracentrifugal measurements.

Similar size and shape data can be obtained from low angle X-ray scattering of dilute solutions. As in the case of light scattering, molecular weights may be obtained from a knowledge of the absolute scattering intensity when the data is extrapolated to infinitely dilute solution state. Further parameters which can be obtained from the so-called Guinier plot when extrapolated to zero concentration, as well as to zero scattering angle, are the radius of gyration of the particle and the persistance length. Assuming certain statistical distribution of the persistance length for the different shapes of molecules, one may calculate theoretical radii of gyration for the different shapes. A comparison between the theoretical and experimental radii of gyration gives an estimate of the shape of the molecule (*180–183*).

In general, one can state that the shape of the acidic polysaccharides depends on the ionic strength of the environment and the pH. At low ionic strength the negative charges of the polymers tend to repel each other and the molecules become extended. This has been demonstrated elegantly during the so-called self-hydrolysis of heparin. The acidic form of heparin catalyzes its own hydrolysis which is primarily a desulfation process. Although the molecular weights change slightly, the axial ratio of the molecular changes drastically as the anionic sites are removed (*184*). At higher ionic strength, when the negative charges on the polymers are largely depressed, the shape of the molecule tends to reflect the intrinsic stiffness of the chains. Obviously the molecular weight also influences the over-all shape. A large molecule with a molecular weight of the order of 10^6 may have large statistical segments and hence large intrinsic stiffness, but since the number of statistical segments is large, the over-all shape of the molecule may be a stiff random coil. This is the case with hyaluronic acid. On the other hand, when smaller molecules show a compact shape, the indication is that they have a large degree of freedom of rotation around the glycosidic linkages. In this respect, it is interesting to note that the animal acidic polysaccharides (chondroitin sulfates heparin, etc.) show a much greater flexibility than acidic polysaccharides from plant origin (pectinic acid, alginic acid, etc.), and this is in spite of the fact that the surface charge density on the animal acidic polysaccharides is usually higher than on those from plant origin. At the present moment there is no clear-cut evidence of what the major influence is in providing intrinsic stiffness to a molecule in solution. Although with neutral polysaccharides it is often cited that the β-glycosidic linkages provide stiffer backbones than the α linkages, mainly on the basis of comparison between cellulosic and amyloidic backbones, this does not seem to work uniformly with polyelectrolytes. Neither can one make a case for the 1-3 glycosidic linkages against the 1-4 nor for the bulkiness of side groups such as *N*-acetyl-*N*-sulfate or *O*-sulfate. Although there are some parameters for intrinsic stiffness which are more accurately measured in the solid state (see Section VI.A), one cannot automatically extrapolate measurements on crystallites to individual molecules in solutions. One reason for this is that the stiffness of a polymer backbone in the crystallite regions may be due more to packing factors, such as the lateral interchain interaction forces, than to steric hindrances to rotation around the glycosidic linkages. Another reason is that the solvation of the single acidic polysaccharide molecule by water molecules is probably the major factor determining shape in infinitely dilute solution.

This solvating power is indicated by the second virial coefficients A_2 of the osmotic pressure and light-scattering measurements. Some of these values are reported in Table I. This second virial coefficient is indicative of the polymer–solvent interaction, and it is dependent on temperature, pH, and ionic strength. A high positive slope of the π/c vs c curves and the zero angle line of the Zimm plot yields a high A_2 value, hence good solvating power, while negative slopes mean "bad" solvents or strong favorable polymer–polymer interaction. As an illustration, one can take the measurements on pectinic acid in $0.155\ M$ NaCl. The negative slope indicates bad solvating power, and it is known that pectinic acids are precipitated with an increase in ionic strength. Hence, the rod shape of the molecules obtained in this solvent (160) may be due more to the solvating properties than to the intrinsic stiffness of the molecules.

In summary, one can conclude that although the physical–chemical measurements in dilute solutions yield important parameters of the single molecules in solution, one cannot and should not take these parameters as representative of physiological conditions. Since the biological role of acidic polysaccharides is generally performed in matrices where we deal with concentrated solutions or gels, the parameters of the single molecule may change drastically due to polymer–polymer interactions. Under physiological conditions the molecular parameters lie probably between those obtained in dilute solutions and in the solid state.

IV. THERMODYNAMICS OF DILUTE SOLUTIONS

A. Thermodynamic Parameters of Mixing

The primary interest of dilute polymeric solution is to study the dependence of chemical potential upon concentration. In general

$$\Delta \bar{G}_M = RT \ln a_1 = -(RT\bar{V}_1/M_2)C_2(1 + A_2C_2 + B_3C_2^2 + \cdots) \quad (9)$$

where a_1 refers to the activity of the solvent, and A_2 and B_3 are the second and third virial coefficients. The equation is quite general for nonelectrolyte solutions as well as for polyelectrolytes (component 2) in the presence of relatively high concentrations of neutral salts (185).

The second virial coefficient is an expression of solute–solvent interaction and if the macromolecular solutions were ideal, its value would be very small. However, the $\Delta \bar{G}_M$ is strongly concentration dependent and the second virial coefficient is usually evaluated by statistical treatment of the excluded volume under limiting conditions where $\Delta \overline{H}_{\mathrm{mix}} = 0$. (See Chapter IV). This does not imply that there is no interaction between solute and solvent, but that we are dealing with such dilute solutions that $\overline{H}_i - \overline{H}_i^\circ = 0$.

General statistical treatments for the second virial coefficients in terms of excluded volume for different shapes can be found in standard textbooks (125,186). Some of the values of the second virial coefficients obtained in acidic polysaccharide solutions are given in Table I. However, since these values are obtained from the slope of osmotic pressure or light-scattering curves intended for molecular weight determination, their significance is mostly qualitative, indicating good or poor solvent power. The second virial coefficient of hyaluronic acid obtained by osmotic-pressure measurements (187) and by sedimentation equilibrium runs (136) agreed well. The ultrafiltration residue of an ox synovial fluid which had high protein content gave a second virial coefficient of 1.8 × 10⁻³, while a deproteinized hyaluronic acid had a value of 2.5 × 10⁻³.

The variation of the second virial coefficient with temperature and ionic strength has rarely been investigated in the case of acidic polysaccharides. For arabic acid the dependence of A_2 on ionic strength was reported (164). As the ionic strength increased from 0 to 16 × 10⁻², the second virial coefficient decreased from 3.2 × 10⁻³ to 0.4 × 10⁻³. Referring to the excluded volume concepts, the authors concluded that the effective volume of arabic acid increased with decreasing ionic strength.

If one studies the temperature dependence of either the second virial coefficient and the physical chemical phenomena from which it is obtained, one gets a better insight into the thermodynamic parameters. For example, the differential free energy of mixing is obtained directly from osmotic pressure

$$\overline{\Delta G} = -\pi \bar{V} \tag{10}$$

The heat of mixing (the subscript 1 refers to the solvent, hence the heat of dilution) can be obtained from the temperature dependence of the free energy, i.e., osmotic pressure

$$\overline{\Delta H}_1 = \frac{\bar{V}_{12}\pi_2/T_2 - \bar{V}_{11}\pi_1/T_1}{1/T_2 - 1/T_1} \tag{11}$$

where $\bar{V}_{12}$, π_2, and T_2 are the partial specific volume, osmotic pressure, and temperature at T_2, the higher temperature, and $\bar{V}_{11}$, π_1, and T_1 the values at the lower temperature.

The entropy can be obtained from

$$\overline{\Delta S}_1 = \frac{\overline{\Delta H}_1 - \overline{\Delta G}_1}{T} \tag{12}$$

These differential thermodynamic functions have been evaluated by Bettelheim for such pectic substances (*188*) as highly esterified pectin, sodium and calcium pectate, and pectic acid.   The differential free energies of dilution decreased with concentration and all have negative values, indicating the spontaneity of the dilution process.   As Eqs. 9 and 10 indicate, the differential free energy, and hence the second virial coefficient, decrease with increasing molecular weight.   The differential enthalpy functions are positive at low concentrations.   For the esterified pectin and the salts of pectic acids, the enthalpy functions pass through maxima at a certain concentration, after which they decrease to negative values. For pectic acid there is only negative enthalpy which decreases with concentration.   This behavior corresponds to the solubility parameters of the different substances.   The differential entropy functions have positive values.   The smallest change in the entropy function with increased concentration is observed with the pectic acid, indicating the most rigid molecular species among the pectic substances follow the order calcium pectate, esterified pectin, and sodium pectate.

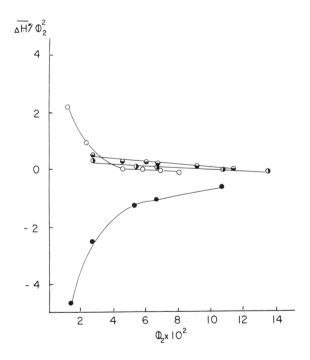

Fig. 12.   Differential heats of dilution of pectin ($\bullet$), sodium pectate ($\ominus$), calcium pectate ($\bigcirc$), and pectic acid ($\bullet$) as a function of composition.

These data can be represented by the so-called excess thermodynamic functions (189) which indicate the deviations from ideality

$$\overline{\Delta G_1^*} = \overline{\Delta G_1} - \overline{\Delta G_1^0} = -A_2(T)C_2^2\bar{V}_1 \tag{13}$$

$$\overline{\Delta H_1^*} = \overline{\Delta H_1} = \left[\frac{dA_2(T)}{dT} - A_2(T)\right]C_2^2\bar{V}_1 \tag{14}$$

$$\overline{\Delta S_1^*} = \overline{\Delta S_1} - \overline{\Delta S_1^0} = \left(\frac{dA_2(T)}{dT}\right)C_2^2\bar{V}_1 \tag{15}$$

If instead of concentration one uses the volume fraction of the polymer, ϕ_2,

$$C_2 = \phi_2\rho_2 \tag{16}$$

In the case of ideal solutions both the $\Delta H_1^*/\phi_2^2$ and $\Delta S_1^*/\phi_2^2$ should be zero; in an athermic solution the $\overline{\Delta H_1^*}/\phi_2^2 = 0$ and $\overline{\Delta S_1^*}/\phi_2^2 = $ constant, while in the "regular" solution of Hildebrand (190) the $\Delta S_1^*/\phi_2^2$ should be zero and $\overline{\Delta H_1^*}/\phi_2^2$ a constant. These excess functions are given in Figs. 12 and 13. They indicate the thermodynamic property deviation from ideality without the need of resorting to a specific model.

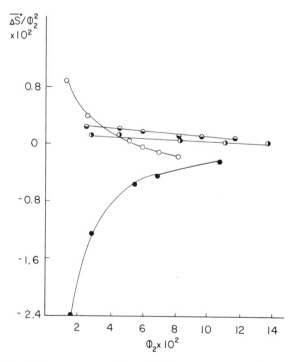

Fig. 13. Differential excess entropies of dilution of pectin (◑), sodium pectate (◕), calcium pectate (○), and pectic acid (●) as a function of composition.

The data could have been treated in similar entropy and enthalpy terms on the basis of one or another better-known solution theories [Flory–Krigbaum (191), Zimm, Stockmayer, and Fixman (192), or Orofino–Flory (193)], but the assumptions involved in the derivations of the thermodynamic parameters in these theories do not provide much greater insight than the excess properties given above which do not require any specific models.

B. Hydrogen Ion Equilibria

The equilibrium which exists between hydrogen ions and acidic polysaccharides have been studied extensively by titration curves. The simplest of these titration curves are those in which all the dissociable sites are identical, such as in polycarboxylic acid. The apparent dissociation constant, K_a, is a function of pH and degree of dissociation, α, and can be described empirically by the Katchalsky–Spitnik equation (194)

$$\text{pH} = \text{p}K_a + n \log\left(\frac{\alpha}{1 - \alpha}\right) \tag{17}$$

According to Spencer (195)

$$\text{p}K_a = n\text{p}K^* \tag{18}$$

and $\text{p}K^*$ is insensitive to changes in ionic strength and concentration while $\text{p}K_a$ and n decrease smoothly with increasing ionic strength and concentration. When two different carboxyl groups are present, two $\text{p}K^*$ values can be obtained with n_1 and n_2 differing from each other.

The dissociation of pectic substances has been studies by many investigators and has been summarized by Deuel, Solms, and Altermatt (196). They point out that while the apparent dissociation constant of the galacturonic acid is 3.25×10^{-4} at 19°C, with pectic substances this apparent dissociation constant varies from $0.1–10 \times 10^{-4}$, but in most cases it is smaller than that of the galacturonic acid. The apparent dissociation constant is generally smaller the higher the surface charge density of the polyelectrolyte, i.e., when the interionic interaction is greater. Hence, these apparent dissociation constants decrease with (1) increasing neutralization of pectin, (2) decreasing pectin concentration, and (3) decreasing esterification of the carboxylic groups of the pectins. The titration curves of pectinic acid and arabic acid closely resemble those obtained with strong monobasic acids. The apparent dissociation constants, however, were concentration dependent. Only at high concentration, 4×10^{-3} equivalent/liter, do the apparent dissociation constants reach a limiting value; at lower concentration they decrease with dilution. The limiting value for K_a for apple pectin was 3.5×10^{-4}, for citrus pectin 7.5×10^{-4}, and for arabic acid 2.5×10^{-4} (*159*).

For pectinic acids (*197*) and pectic acids (*198*) it was found that the pK_a varied almost linearly with the square of the carboxyl ion concentration. The zero concentration extrapolated value of pK_a for each of these materials was 2.7, which the authors (*197*) took for the fundamental dissociation of a simple carboxyl group when no adjacent groups are ionized.

However, it was found that the empirical Eq. 17 can describe the dissociation of pectinic acids with different degrees of esterification (*199*). At 60% esterification the $pK_a = 3.8$ and $n = 1.2$, and both increased with decreasing degree of esterification.

Similarly it was found that carboxymethyl cellulose has pK_a values which increase with the degree of substitution, i.e., for D.S. $= 0.55$, $pK_a = 3.44$; D.S. $= 0.65$, $pK_a = 3.65$; D.S. $= 0.72$, $pK_a = 4.20$ (*200*).

For carboxymethylamylose the pK_a was found to be 3.43 while the empirical n was 3.60 in water and 1.1 in $0.5 M$ KCl solution (*201*). Alginic acid similarly obeys the empirical Eq. 17. Lately, however, alginic acids proved to be heterogeneous, i.e., besides being anhydro-β-D-mannuronic acid polymers, they also contained guluronic acids. The variation in pK_a values obtained for different preparations reflected the composition: pK_a for guluronic acid is 3.65 and for mannuronic acid 3.30 (*202*).

To obtain the intrinsic ionization constant, one has to include the electrostatic potential Ψ_0 of the polyion in the $pK_0 \sim pH$ relationship (*203*)

$$pH = pK_0 - \log \frac{1 - \alpha}{\alpha} + 0.4343 \frac{\epsilon \Psi_0}{kT} \tag{19}$$

This potential has been related to the differential electrostatic energy for a highly stretched polyion (*204*) by

$$\Psi_0 = \frac{1}{\epsilon} \frac{\delta Fe}{\delta v} = \frac{2\alpha\epsilon}{Dbj} \ln \left(1 + \frac{6}{\chi Sb} \right) \tag{20}$$

where j is the distance between COO^- groups in terms of monomeric units, b length of the repeating unit, χ is the inverse Debye radius $= (0.327 \times 10^8 I^{1/2})$ where I is ionic strength, S is the number of monomers in a statistical segment, and D is dielectric constant.

Mathews (*148*) applied Eqs. 19 and 20 to chondroitin-4-sulfate and dermatan sulfate by replacing α with $1 + \alpha$ since in the titration range of the carboxylic groups the sulfate groups are fully ionized. He found that chondroitin-4-sulfate has a pK_a value of 3.38 and $n = 1.32$ from Eq. 17 and a pK_0 value of 3.06 from Eqs. 19 and 20. Dermatan sulfate on the other hand has $pK_a = 3.82$, $n = 1.25$, and $pK_0 = 3.49$.   When Mathews compared the calculated Ψ_0 values to those obtained from electrophoretic mobilities, he found that the experimental values were twice as large as those calculated.   These experimental electrophoretic potentials would yield a pK_0 value about 0.3 of a unit lower than those calculated.   In any case, the difference between chondroitin-4-sulfate and dermatan sulf- ate is real in pK_0 values.   The pK_0 value of hyaluronic acid is 3.20 when extrapolated to $\alpha = 0$ and 3.30 when using Eqs. 19 and 20.   Mathews interpreted the different pK_0 values for chondroitin-4-sulfate and dermatan sulfate as the different interactions between glucuronic carboxyl and sulfate groups, and iduronic carboxyl group and sulfate groups.

Titration curves have been used to differentiate between protons from sulfate and carboxyl groups in heparin (205).   However, for chondroitin sulfates this is not possible in aqueous solutions.   Combined titration curves in water and in aqueous dioxane mixtures were recently used to estimate the sulfate and carboxyl protons (206).

C. Binding of Cations

Rice and Nagasawa (207) distinguish three types of ion binding by polyelectrolytes: (1) Undissociated binding, i.e., —COOH which was dealt with in the previous section; (2) ion pair formation of localization of an ion in the region of one or a few charged sites; or (3) nonlocalized binding or retention of an ion in the potential field of the polyelectrolyte. All three types may occur simultaneously, but at present we are interested only in the last two types.

If one assumes that the binding can be treated in terms of multiple equilibria and, as a first approximation, one can neglect lateral interac- tions among the bound ions, Klotz's expression can be used (208)

$$\frac{1}{r} = \frac{1}{nk(A)} + \frac{1}{n} \tag{21}$$

$$\frac{r}{(A)} = kn - kr \tag{22}$$

where r is the number of moles of bound cation per total repeating periods of the acidic polysaccharides, n the binding sites per repeating period, k the association constant, and (A) is the free cation concentration.

Mathews has investigated the nature of binding of $Co(NH_3)_6^{6+}$ to differ- ent acidic polysaccharides of the connective tissues (209).   Equation 21 was used to evaluate association constants.   When sufficient $HClO_4$ was added to the solutions to suppress the ionization of the carboxyl groups, chondroitin-4-sulfate, chondroitin-6-sulfate, and dermatan sulfate bound the cation to the same extent as the keratosulfate which has only one anionic (sulfate) group. When the ionization of the carboxyl group was not suppressed, the binding of cation was strongest with heparin and in decreasing order with dermatan sulfate > chondroitin-4-sulfate > chondroitin-6-sulfate > keratosulfate. When desulfated chondroitin sulfates and dermatan sulfates were used, the desulfated dermatan sulfate bound the cations best; the desulfated chondroitin sulfates had about half of the association constant of dermatan sulfate, and keratosulfate bound the $Co(NH_3)_6^{3+}$ the weakest. Mathews concluded that the higher binding capacity of dermatan sulfate is due to the fact that both the carboxyl and sulfate groups are axial. In chondroitin-4-sulfate only the sulfate group is axial, and in chondroitin-6-sulfate neither of the anionic groups is axial. The lowest degree of ion pair formation is with keratosulfate which contains one charge per period and has an intercharge distance of about 10 Å. Heparin, with 3.5 charges per period, has the highest surface charge density; hence it provides the best ion pair binding.

A similar trend was shown in the binding of Ca^{2+} by mucopolysaccharides under physiological conditions (210).

The binding constants or association constants of carboxylated starch were obtainable for K^+, Na^+, and Methylene Blue from Eqs. 21 and 22. The Methylene Blue required two carboxylic sites at 0.01 ionic strength but the data extrapolated to zero ionic strength indicated one carboxylic group per Methylene Blue (211). The association constants of Na^+ and K^+ were two magnitudes smaller than that of the Methylene Blue. The enthalpy and entropy of binding were both positive for Methylene Blue as well as for K^+ and Na^+. The positive entropy values indicate that water of hydration is released both from the polyelectrolytes and from the small counterions, and that the configuration of the extended polyanion chain is changing to a random coil as ion pair formations occur. The $\Delta S°$ was 40 e.u. for Methylene Blue and 55 e.u. for Na^+ and K^+ calculated for a mole of COO^- in the starch.

Balazs et al. have shown that the interaction of polyanions with cationic dyes over a large range of concentration cannot be expressed by Eqs. 21 and 22.

Multiple equilibria is suggested only when the dye-to-anionic site ratio increased. However, under such conditions large lateral interaction occurs between the stacked up cationic dyes (212). The equilibrium constant, $K_{c'}$ between heparin and azur A was obtained from the Benesi–Hildebrand equation (213):

$$\frac{C_D C_s}{A - A_0} = \frac{1}{K_c L (E_{DS} - E_D)} + \frac{C_s}{(E_{DS} - E_D)L} \qquad (22a)$$

where C_s is the equivalent initial concentration of anionic sites of heparin;

C_D is the initial molar concentration of the dye; E_D and E_{DS} are the molar absorbances of the dye and the complex, respectively; A_0 is the absorption of the initial concentration of the dye; A is the total absorbance of the mixture at the wavelength under consideration; and L is the length of light path. From the equilibrium constant and its temperature dependence, the following thermodynamic parameters were obtained at 25°C.: $\Delta G = -4.65$ kcal/mole; $\Delta H = -7.3$ kcal/mole; $\Delta S = -8.8$ cal/mole degree. The complex thus formed, and similarly many others between cationic dyes and polyanions, are paramagnetic when exposed to light. Such photo-induced paramagnetism is due to charge or electron transfer (214).

A new and sensitive method described by Balazs et al. (215) can be used to study polyanion–cationic dye interactions. This involves the rate of reaction of hydrated electron with cationic dyes. The fast rate of reaction is decreased in the presence of polyanions and restored under conditions which usually inhibit metachromesia.

The 1:1 complexing between anionic sites and cationic dyes was used by Stone and Bradley to titrate spectrophotometrically a large variety of acidic polysaccharides. Acridine Orange was found to be applicable for quantitative analysis for all kinds of polyanions. Methylene Blue was applicable to acidic polysaccharides which had at least one anionic site per sugar ring and in which the negative charges occurred in a regular sequence. Titration curves with Methylene Blue and proflavine provided a spectrophotometric means of distinguishing between relatively homogeneous and nonhomogeneous charge distribution in acidic polysaccharides (216). The spectroscopic characteristics of the interaction between anionic sites and cationic dyes, of the so-called metachromatic reaction, are loss of intensity and shift in frequency. These two properties were described quantitatively in terms of optical parameters such as oscillator strength and transition moment vector. The optical parameters were correlated with general metachromatic behavior for a large number of systems (217).

In studying the binding of cationic dyes to acidic polysaccharide sites, Scott (218) used the critical electrolyte concentration (CEC) as an indicator of the affinities of the cations. If R^+ is the precipitating dye cation and M^+ is the counterion, the relative affinities of these two types of cations for the anionic sites determine the CEC. The CEC is the smallest with $COO^- < =PO_4^- < -SO_4^-$ when R^+ is an ion of low electrostatic field and polarizing power. The sequence is the opposite if R^+ contains a chelated Al^{3+} with high polarizing power.

The relative affinities of different cations for the same anionic sites were evaluated by Dunstone (219) for chondroitin-4-sulfate, dermatan sulfate, heparin, and heparin monosulfate by calculating exchange parameters with a variety of techniques. The order of increasing cation infinities were:

$$K^+ < Na^+ < Mg^{2+} \leqslant Ca^{2+} \leqslant Sr^{2+} \leqslant Ba^{2+}$$

This suggested that the ion binding is nonspecific electrostatic binding rather than the chelating type where the order should be

$$Mg^{2+} > Ca^{2+} > Sr^{2+} > Ba^{2+}$$

For alginic acid the affinity of cations was in the following order (*220*):

$$Na^+ < Co^{2+} < Ca^{2+} < Ba^{2+} < Cu^{2+}$$

and for pectic substances (*196*)

$$NH^+ < Li^+ < Na^+ < Ca^{2+} < Ba^{2+}$$

and for carboxymethyl cellulose (*221*)

$$Li^+ < Na^+ < K^+ < NH_4^+ < Mg^{2+} < Ca^{2+} < Sr^{2+} < Ba^{2+} < Zu^{2+} < Cu^{2+}$$

A special selectivity was found in alginic between binding Ca^{2+} and Sr^{2+}. Alginic acid with high guluronic content binds Sr^{2+} more strongly than those with high mannuronic acid content. Alginic acid containing 80% guluronic acid had a Sr^{2+}/Ca^{2+} selectivity coefficient of 5. This has also been proven clinically in rats and humans, where alginate inhibits strontium uptake (*222*).

The preferential binding of K^+ vs Na^+ ion in the connective tissues by acidic polysaccharides gave rise to much speculation. In heparin the selectivity coefficient of K^+ over Na^+ is 1.40 between pH 7–12 and concentrations of 2–25 g/liter (*223*). In chondroitin sulfate no such selectivity was observed (*224*). The explanation is that with anionic groups such as $=PO_4^-$ and COO^- that are more polarizable than water, the cation binding is determined by its naked or unhydrated size. Hence Na^+ is bound more strongly than K^+ since the Na^+ is smaller. When the polarizability of the anionic site is less than that of water ($—SO_4^-$) the hydrated size of the cation is the determining factor. Hence K^+ will be bound preferentially over Na^+. Since heparin has more sulfate groups than COO^-, its selectivity of K^+ is explainable; chondroitin sulfate, having equimolar SO_4^- and COO^- groups, shows no such selectivity. This qualitative rule is not obeyed universally (see *221*) and even if it were, the small selectivity coefficients of acidic polysaccharides would not account for the large selectivity of the tissues.

V. TRANSPORT PHENOMENA

In the previous section physical chemical properties were discussed for systems at equilibrium, and hence the laws of thermodynamics were applicable. When we deal with transport phenomena, we usually talk about the flux or flow of matter per unit area and the systems are not at equilibrium. The thermodynamics of irreversible processes describes these phenomena under the condition that one is not too far removed from equilibrium. If one designs the experiment in such a manner that unidirectional flow results and all the forces act along this direction and are relatively small, the so-called phenomenological equations of irreversible thermodynamics can be applied. These state that the flux is a linear function of the forces which causes the flow

$$J_i = \sum_k L_{ik} F_k \tag{23}$$

where J_i is the flux per unit area across a plane perpendicular to the x direction of compound i, F_k are the forces acting along x, L_{ik} are the phenomenological coefficients which, like thermodynamic variables, are functions of temperature, pressure, and composition (225). However, to evaluate the phenomenological coefficients one has to resort to mechanical models of the transport phenomena. The mechanical model which describes most of the acidic polysaccharides in transport processes is the free draining coil with a certain degree of stiffness, as indicated by the length of the statistical segment. The concept of such a hydrodynamic solvated "particle" has been developed and described in Chapter II. To describe the thermodynamically excluded volume of the polymer molecule, one can resort to the concept of an equivalent hydrodynamic sphere which we visualize as a solid sphere with radius R_e. This is related to the radius of gyration, R_g, of the random coil obtained from light-scattering measurements by a constant, ζ, which certain solution theories predict as being a universal constant (186).

$$R_e = \zeta R_g \tag{24}$$

A. Viscosity

Among the fluxes which results in a translational motion of macromolecules, the best known is the viscosity. A moving particle under the influence of an applied force encounters an opposing frictional force in viscous fluids

$$ma = m\left(\frac{du}{dt}\right) = F - fu \tag{25}$$

where m is the mass of the particle; $a = du/dt$, the acceleration; and u the velocity. For viscosity experiments in a steady state the velocity will be maintained and this velocity is

$$u = \frac{F}{f} \qquad (26)$$

where F is the driving force and f the frictional coefficient. Hence the flux in particles/sec/cm²

$$J_i = \frac{N_i F}{f_i} \qquad (27)$$

where N_i is the number of particles/cm³ and, comparing this with Eq. 23,

$$L_{ik} = \frac{N_i}{f_i} \qquad (28)$$

The phenomenological coefficient can be obtained if we use a reasonable hydrodynamic model to obtain the frictional coefficient in terms of viscosity coefficient and radius of gyration of the particles.

If we take the simplified mechanical model of equivalent hydrodynamic sphere, the frictional coefficient can be obtained from Stoke's law

$$f = 6\pi\eta R_e \qquad (29)$$

The viscosity of dilute polymer solutions and its concentration dependence is described by the Huggins' equation

$$\eta_{sp}/c = [\eta] + k[\eta]^2 c \qquad (30)$$

where η_{sp}/c is the reduced viscosity, $[\eta]$ the intrinsic viscosity, c the concentration and k the Huggins' constant indicating solvent–solute interaction.

The reduced viscosity is also shear rate dependent since asymmetric molecules like acidic polysaccharides usually orient at high velocity gradients which leads to a reduction of the reduced viscosity with increasing shear rate. This was demonstrated with sodium carboxymethyl cellulose. A distinct change in the intrinsic viscosities with shear rate was observed between 1000–5000 sec⁻¹ (226). On the other hand, at low shear rates, 50–1000 sec⁻¹, no significant change in intrinsic viscosity could be found on a comparable carboxymethyl cellulose sample (227).

A more recent reinvestigation of the same sample in 0.01, 0.1, and 0.2 N NaCl solutions indicated that the intrinsic viscosity dependence on the shear rate over the whole range could be described by a hyperbolic equation

$$[\eta]_q = [\eta]_\infty + \frac{a}{b + \beta} \qquad (31)$$

where β is the shear rate and a and b are constants. Moreover, the Huggins' constant (Eq. 30) first decreased with increasing shear rate up to

1000 sec^{-1} and then reached a flat minimum.   This is in contrast with nonpolyelectrolytes in which case the k decreases with increasing shear rate to zero (228).

A modified Huggins' equation was described by Shurz and Pippen (229) in which the shear rate dependence is indicated by

$$(\eta_{\mathrm{sp}}/c)_\beta = [\eta]_0 - a\rho^2 + (k[\eta]_0^2 - b\beta)c \qquad (32)$$

where a is a parameter of the single molecule and b is a parameter of intermolecular interaction, both shear rate dependent. For sodium hyaluronate in water the $[\eta]_0 = 21$, $k = 0.04$, $a = 1.85$, and $b = 1.1 \times 10^{-2}$.

Hence with very viscous material the intrinsic viscosity has to be obtained not only by extrapolation to zero concentration but also to zero shear gradient.   The intrinsic viscosity thus obtained has a direct relationship to the molecular weight.   For a random coil this relationship is

$$[\eta] = \Phi(\overline{r^2})^{3/2}M \qquad (33)$$

as the Flory–Fox equation (230) or

$$[\eta] = 10/3\pi N\zeta^3 R_g^2/M \qquad (34)$$

in Tanford's nomenclature (186).

In both cases a universal constant Φ or ζ was predicted by the theory, and since in Eq. 34 all parameters except ζ are obtainable experimentally, intrinsic viscosity values combined with light-scattering data were used to evaluate ζ.   For sufficiently large polymers this was 0.875, but ζ was not a constant for smaller molecular weights.   Therefore, the universality of this coefficient is in doubt even for random-coil models.

However, acidic polysaccharides are not always free draining random coils but sometimes stiff rods or rigid random coils.   The empirical Staudinger equation relating intrinsic viscosity to molecular weight

$$[\eta] = K_m M^a \qquad (35)$$

has two parameters: a is the shape factor (0.75 for random coil, 0.5 for compact sphere, and 1 and greater for extended rigid rod). K_m is a solvent–solute interaction parameter.

Intrinsic viscosities and K_m and a values of some acidic polysaccharides are listed in Table II.   One can see that animal acidic polysaccharides such as hyaluronate and chondroitin sulfate have values close to 0.75, indicating a random-coil shape, while pectinic acid, alginic acid, agar-agar, and carboxymethyl cellulose have a rigid-rod structure.

However, as one sees in the case of carboxymethyl cellulose, shape is dependent on the ionic strength of the media.   Viscosity measurements on

TABLE II[a]

Hydrodynamic Parameters of Acidic Polysaccharides

Substance	Source	Medium	$S_w^{20} \times 10^{13}$	D^0, cm²/sec	$\bar{V}$, ml/g	μ, cm²/V-sec	$[\eta]$, dl/g	K_m	a	Axial ratio	Ref.
Hyaluronic acid	Vitreous	0.006PO₄	2.8	2.31	0.66						141
	Body	0.1 N NaCl					2.3				142
		0.2 N NaCl	2.46–7.15	0.34–1.74			2.5–24.5	0.036	0.78	70	133
	Umbilical cord						2.3–7.0				143
							9–54				143
	Synovial fluid	0.2 M PO₄				8.0	31–47				175
							10–55				143
	Normal						39.3				177
											178
	Pathological						24–30				179
	Streptococcus	0.01 M PO₄				6.0	1.75				147
Chondroitin-4-sulfate	Bovine nasal septa	0.01 M PO₄				9.5					147
		0.1 Barbital				13.2					231
		0.15 M PO₄ 0.2 M NaCl					0.51–1.1	3.1×10^{-4}	0.74		148
Chondroitin-6-sulfate	Chondroma	0.15 M PO₄ + 0.2 M NaCl					0.84				148
Dermatan sulfate	Beef lung	0.15 M PO₄ + 0.2 M NaCl					0.34–0.57	3.1×10^{-4}	0.74		148
	Pig skin	0.2 M NaCl			0.57						150
	Cornea	0.5 M NaCl	0.96–1.72	4.1–6.9	0.47–0.55		0.13–0.55				151
Keratan sulfate							0.23				147
Heparin	Hog	Tris B				5.5 and	12.5				232
	Hog lung		2.10	8.51							153
	Human lung		2.27	6.29		11–18					153
		0.1–1.0 N NaCl		7.70	0.47	14–22					154
	Hog-beef	0.02–0.3PO₄				18–19					154
		0.2 M NaCl	2.13	7.45	0.42		0.12–0.19	1.58×10^{-2}	1.00		166

Polysaccharide	Source	Solvent	s_{20}	D^0	$\bar{V}$	μ	$[\eta]$	MW	K_m	a	Ref.
Pectinic acid	Citrus	0.155 M NaCl	4.0	2.0	0.612	9.9	1.2–8.7	53–165	1.4×10^{-6}	1.34	*160*
Alginic acid	Citrus	0.1 M NaCl					5–8		8×10^{-5}	1.00	*159*
		0.2 M NaCl									*233*
		0.2 M NaNO₃	2.4	3.0	0.605						*159*
		0.1 M NaCl									
Carrageenan			3.2–6.7	1.4–0.6			13.0	160–340	2.4×10^{-5}	1.00	*161*
Agar–agar						12	3.4–11.2		3.7×10^{-4}	1.00	*162*
Arabic acid	Gum arabic	0.2 M NaCl	10.4	2.2	0.32	20	2.9–16.1		1.3×10^{-2}	0.54	*163*
	Gum arabic	1 N NaCl					0.12				*159*
	Gum senegal	1 N NaCl					0.16				*90*
	Gum nubica	1 N NaCl					0.10				*91*
Pneumococcal polysaccharide	Type I	0.15 M	6.5	2.0				60			*90*
	II	NaCl	7.2	0.75				200			*90*
	III		4.3	1.00				110			*166*
Techoic acid	Staphylococcus	Native acid ext.	1.66	8							*167*
			0.74	14							
Glucan sulfate	B. undatum	0.5 M NaCl	9.75	3.28	0.6		5.0				*68*
	C. lampas		15								*65*
Carboxymethyl cellulose		0.2 M NaCl	5.80		0.55		11.75		4.3×10^{-4}	0.74	*168*
		0.005 M NaCl	3.97				34.8		7.2×10^{-5}	0.95	*169*
		0.005–0.5 M NaCl					6.8–20.5				
Amylose sulfate		0.1 N NaCl					1.10				*171*
Amylopectin sulfate		0.1 N NaCl					0.94				*171*
Guaran sulfate		0.1 N NaCl					4.90				*171*
Dextran sulfate	Linear	0.05–0.2 M NaCl				11.4–17.8	0.164				*172*
	Branched						0.157				*173*
Carboxy-polyglucose							0.6				*173*
											123

a s^{20} = sedimentation coefficient; D^0 diffusion coefficient; $\bar{V}$ = partial specific volume, μ = electrophoretic mobility; $[\eta]$ = intrinsic viscosity; K_m = solute solvent parameter of Eq. 35; a = shape factor of Eq. 35.

polyelectrolytes are usually performed in the presence of neutral salt. This is done in order to avoid the electroviscous effect which demonstrates itself in a rapidly rising reduced viscosity with increased dilution and makes the extrapolation to zero concentration difficult. The increase in reduced viscosity with dilution is due to the greater ionization at low concentration and hence a more extended rigid-rod shape due to mutual repulsion of the anionic sites. In the presence of neutral salt this ionization is greatly reduced. Therefore the degree of ionization in the presence of salt will influence the over-all shape of the molecule as can be seen with carboxymethyl cellulose which seemingly has a random-coil shape in 0.2 M NaCl and a rigid-rod shape in 0.005 M NaCl solution (168). The effect of charge density on the shape of the molecule is evident in the case of heparin when compared with chondroitin sulfate. Heparin assumes a rigid-rod shape (156) while the less highly charged chondroitin-4-sulfate is a random coil (148).

The over-all shape is the combination of the effect of charge density and the intrinsic stiffness of the polymer backbone. This can be seen clearly with dextran sulfate (173) for which the intrinsic viscosities increased with increasing sulfation. However, at the same degree of sulfation the reduced viscosity of the branched polymer was greater than the linear polymer at high concentration, while at low concentrations the reverse was true. The authors' explanation of this unexpected phenomena was that the branched polymer had a higher charge density; hence greater expansion occurred due to mutual repulsion of the sulfate group. On the other hand, at high dilution the linear chain can expand even further; hence the reduced viscosity of this species will increase to a greater extent with dilution.

The viscosity behavior of solutions of Acacia gums from different species shows that, in accord with the Staudinger expression, the molecular shape is a more important influence than the molecular weight. Molecules from *A. senegal* gum showed greater viscosity than those from *A. arabica* and *A. nubica* in spite of the fact that the latter had higher molecular weights (see Tables I and II). This has been explained as being due to the gum senegal molecules having an open-branched structure compared to the compact-branched structure of the other two species (234,235).

In the case of pectinic acids the reduced viscosity decreased with decreasing pH as expected because the acidic medium suppressed the ionization of the carboxyl groups. The reduced viscosity, on the other hand, increased with increasing esterification of the carboxyl group. Deuel et al. (196) explained this as a steric effect; namely, that the introduction of side chains hinders the rotation around the glycosidic bond and extends the molecule.

In dilute aqueous alginic solutions, 0.002–0.1%, a typical polyelectrolyte behavior was observed in the upward curvature of the η_{sp}/c vs c graph with increasing dilution. However at high concentrations, from 0.1–1% solution, a positive slope was obtained with a resulting minima in the intermediate concentration range. The authors (233) explained this behavior by the relative stiffness of the alginate molecule. The concentration dependence of viscosity is due to two factors: (1) At low concentration the degree of ionization increases with dilution and the molecule is expanding with dilution, and (2) intermolecular interactions increase with concentration and increase the viscosity. If the molecule was flexible, the increase in intermolecular interaction with increase in concentration would be counteracted by the contraction of the molecule and no minimum would appear in the η_{sp}/c vs c curves. But with the stiff alginate molecule no appreciable contraction of the molecular size occurs beyond a certain concentration range, hence the intermolecular interaction will be the important factor at high concentrations.

B. Diffusion–Sedimentation

Diffusion occurs whenever there is a concentration gradient in a system which otherwise is at equilibrium. In a one-dimensional diffusion of a two-component system Fick's first law is applicable

$$J = -D\left(\frac{\partial C}{\partial X}\right)_t \tag{36}$$

where J is the flux, D is the diffusion coefficient, and $(\partial C/\partial X)_t$ is the concentration gradient.

In the case of acidic polysaccharides the experiments usually are performed in aqueous salt solutions, and hence the situation is more complicated. First there is a three-component system, and second the concentration gradient is influenced by the electrostatic potential. However, these difficulties are usually neglected because of the argument that the third component (salt) is uniformly distributed initially and, if it does not interact with the polyanion, its flow will be minimal. Furthermore, the high salt concentration suppresses any appreciable amount of ionization of the macromolecules, hence it minimizes the appearance of electrostatic potential. Diffusion coefficients are usually obtained by measuring the concentration gradient by the refractive index gradient at different times, t.

The area and the height of the refractive index gradient curve is then

$$A/h_{\max} = (4\pi Dt)^{1/2} \tag{37}$$

The diffusion coefficient is concentration dependent and at infinite dilution, D^0, it is inversly proportional to the frictional coefficient

$$D^0 = kT/f \tag{38}$$

A particle sedimenting in an ultracentrifuge has a flow velocity, u, which depends on the rotor speed, ω, and on the distance from the center of rotation, r. The sedimentation coefficient can be defined by molecular parameters

$$S = \frac{U}{\omega^2 r} = \frac{M(1 - \bar{V}_2\rho)}{Nf} \tag{39}$$

Hence the flux in such an experiment is determined by both the diffusion and sedimentation coefficients

$$J = sC\omega^2 r - D\left(\frac{\partial C}{\partial r}\right)_t \tag{40}$$

The sedimentation coefficient is also concentration dependent, and for most acidic polysaccharides one can obtain the sedimentation coefficients at infinitely dilute solutions by extrapolation from the equation

$$\frac{1}{S} = \frac{1}{S_0} + kC \tag{41}$$

The electrostatic potential on the macroions has a great effect on the sedimentation coefficient. The primary charge effect which arises when the macroions sediments away from its counterions may decrease the sedimentation velocity. However, this can be eliminated by the addition of neutral electrolytes. Care must be taken that the added electrolyte does not bind to the macroion and that the sizes of the ions in the neutral electrolyte are the same or a secondary charge effect will arise when the ions of the added electrolyte sediment away from their counterions (236). Sedimentation and diffusion coefficients of acidic polysaccharides are given in Table II.

Varga (145) found that the sedimentation coefficient of hyaluronic acid decreased with a decrease in ionic strength. This could be interpreted in terms of change in the effective charge on the macromolecules. From electrophoretic and sedimentation data, he calculated that at 0.12 ionic strength only 14% of the total charges are effective, while at 0.02 ionic strength 80% were effective. Laurent et al. (133) studied the concentration dependence of the sedimentation and diffusion coefficients for different hyaluronic acid fractions. The sharpness of the boundaries in sedimentation increased with concentration and molecular weight. The reciprocal of the sedimentation coefficients varied directly and linearly with concentration for lower molecular weight fractions (below 10^6) and as a hyperbolic function above a molecular weight of a million. However, the sedimentation velocity was independent of molecular weight at concentrations above 2 g/liter. The diffusion showed a linear relationship with concentration in each case, and the concentration dependence was found to increase with molecular weight. The diffusion patterns showed different skewness for different fractions. The skew ratio, which is the ratio of the areas on both sides of the maximum of the diffusion pattern, increased with concentration and molecular weight. However, this skew ratio extrapolates to 1 at infinitely dilute solutions.

Strong non-Gaussian patterns of diffusion were also reported by Säverborn (159) with arabic acid. The diffusion patterns were steep toward the solvent and this "front" moved rapidly, causing a division of the substance into two concentration ranges with two maxima, one of which corresponded to the original boundary.

C. Electrophoresis

If an electrical field is acting upon macroions, a flow will result in which the velocity of the movement, v, is assumed to be proportional to the electrical field strength. The electrophoretic mobility, u, is the velocity per unit field strength, E,

$$u = v/E \tag{42}$$

From the point of view of the thermodynamics of irreversible processes the phenomenological equation for electrophoresis should be

$$J = L_0 E + \sum L_i \left(\frac{\partial \mu}{\partial x} \right)_t \tag{43}$$

where $(\partial \mu / \partial x)$ represents the concentration gradient of each species in solution and L_0 and L_i are the phenomenological coefficients. Most existing electrophoretic theories neglect the second term in Eq. 43, and probably the contribution of the chemical potential is not an important one.

The molecular parameter which influences the electrophoretic mobility is the surface charge density of the polymer. In the equations most frequently quoted for electrophoretic mobilities, the zeta potential, ζ, appears as proportional to the mobility

$$u = \frac{D\zeta}{C\eta} \tag{44}$$

D is the dielectric constant of the medium and η is the viscosity coefficient. C is a variable parameter for different models, i.e., if the polymer is large compared to the electrical double layer, C is 4 (Smoluchowski's equation) (237). In Henry's equation $C = 6$ (238) is applicable for spherical ions of any size. For other models, such as a cylinder, the value may fall between these two extremes. Acidic polysaccharides can be described by many models. For example, in the electrophoresis of hyaluronic acid, chondroitin-4-sulfate, and heparin, Foster and Pearce (239) found that Smoluchowski's equation was applicable up to 3 V/cm field strength and up to 16 hr. The electrophoresis of acidic polysaccharides is done in an excess of buffer ion concentration so that the specific conductance of the solution is almost entirely due to the small ions present. For this purpose the acidic polysaccharide solutions are dialyzed against the buffer before electrophoresis. The mobility is measured in the descending boundary in spite of the fact that the ascending boundary is sharper. The theoretical reasons for this have been reviewed by Longsworth (240). Some electrophoretic mobilities for the acidic polysaccharides are given in Table II.

The study of mobility vs pH has been performed on acidic polysaccharides. For chondroitin sulfate (231), alginic acid (241), and pectin (242,243) an increase in mobility generally accompanied the ionization of the polymer's acidic groups, and the greater the charge on the colloid surface, the faster it migrated. Douglas and Shaw (241) studied alginic acid and agar-agar, which had been adsorbed on Nujol droplets, as a function of pH. The medium they selected prohibited any polysaccharides lacking acidic or basic groups to acquire charge by ionic adsorption. The mobility and charge on the polysaccharides were therefore completely dependent on their ionizable groups. For agar-agar the mobility showed very little decrease even at a very low pH because the surface charge is due to ionization of strongly acid–SO_4H groups. For alginic acid the increase in mobility over the pH range 2 to 5 was consistent with the behavior of the COO^- groups in this polymer. The change in mobility at pH 6, however, corresponded to only 1 effective COO^- per 10 sugar units and not 1 per unit as the primary structure would indicate.

In chondroitin-4-sulfate the mobility decreased about 30% from pH 5 to pH 2. If one attributes this to the protonation of the carboxyl groups, one would expect a 50% decrease. Warner and Schubert (231) state that this behavior is consistent with the finding of an extensive association of counterions with polyelectrolytes in solution. The number of cations associated may be around 50 to 70% of the number of negative sites available on the polyanion. As the number of sites is reduced, as in the protonation of the carboxyl groups on chondroitin sulfate, the number of associated counterions is *more* than proportionally reduced. Therefore the reduction of net charge is less than the drop expected from protonation.

This association of counterion with polyelectrolyte is very significant. For example, if sodium ions in a buffer are replaced by calcium ions at constant ionic strength, the mobility of chondroitin sulfate is reduced. This was interpreted as the higher association of the calcium over the sodium ions (231).

Associations and aggregations, often by physical entanglement, may sometimes result in obtaining one moving boundary when clearly a mixture of acidic polysaccharides are present. The chondroitin-4-sulfate is a typical case. When isolated in the nondegraded form from bovine cartilage, the acidic polysaccharide is largely bound to protein by covalent bonds although some single-chain chondroitin sulfates with minimal amino acid residues are also present. Electrophoresis cannot separate these fractions although they have distinct mobilities once they are separated by other means such as gel filtration. The only time electrophoresis can achieve separation is at very alkaline pH's, above 12.5, where most of the acidic polysaccharide chains cleave irreversibly off the protein (244).

It has been shown by Bettelheim (188) that borate complexes of neutral polysaccharides, such as galactan, and of polygalacturonate have distinct mobilities; but when they are mixed a single moving boundary with a mobility proportional to the weight fraction of the species present is obtained.

The electrophoretic mobility of pectinic acid has been shown to depend on the degree of esterification. As expected, the mobility is inversely proportional to esterification; at 100% esterification no acid carboxyl groups remain and the mobility is zero, but the relationship was not linear as one would have expected. Ward, Swenson, and Owen (243) explained that sodium ions under the influence of the highly ionized sodium buffer salts are closely associated with the pectate polyion. This results in a decrease in charge on the pectate ion and a resulting decrease in mobility.

An interesting case is the electrophoretic mobility of heparin. Wolfrom and Rice reported that two peaks were obtained when heparin was subject to electrophoresis, and that the first peak contained all the biologically active material (245). Barlow et al. (153,154,232) found that the two peaks were both biologically active. They believed both peaks to be heparin since every heparin fraction showed the same two peaks in the same ratio when subjected to electrophoresis. They suggested that heparin interacts with the buffer ions, leading to complexes. As ionic strength of the medium increased, the faster peak disappeared, which they took as a strong indication that a boundary reaction was occurring.

An interesting aspect of complex formation on electrophoretic mobility has been reported by Niedermeier and Gramling (246). They found that hyaluronic acid isolated from synovial fluid with a relatively high protein content had mobilities which were concentration dependent. The unique aspect of this observation was that as the concentration increased the mobilities increased. If an increase in concentration causes association by entanglement, one would expect a decrease in mobilities with increasing concentration. Bettelheim has suggested that the above observation can be explained if the covalently linked residual protein on the hyaluronic acid has a somewhat hydrophobic character. Under such condition the increase in concentration will cause an aggregation—the aggregating sites being the protein moieties. Since the protein moiety has a smaller surface charge density than the acidic polysaccharide, such an aggregation, in essence, would decrease that part of the surface which has the smaller surface charge density. The net effect will be an increase in the surface charge density of the aggregate as compared to the single molecule with the resultant increase in the mobility. That this hypothesis may be correct has been shown by water vapor sorption studies of Block (247). Hyaluronic acid from synovial fluid had nonreproducible water vapor sorption isotherms with successively less and less sorptive capacity. This was in contrast with other hyaluronic acid preparations, even with those of high protein content (from umbilical cords) which showed good reproducibility or change in sorptive capacity which increased with successive hydration isotherms. On the basis of analogy with other nonreproducible and decreasing sorptive capacities of successive isotherms, it has been proposed that the mechanism responsible for this behavior is due to the conformational changes in the protein moiety, indicating a high degree of hydrophobicity.

D. Flow Birefringence

In the previous three sections the physical chemical measurements characterized translational transport processes. Flow birefringence can give information on the rotational motion of the molecules, and in analogy with Fick's law the phenomenological equation can be written

$$J(\phi) = -\theta[\partial \varphi(\phi)/\partial \phi]_t \tag{45}$$

where θ is the rotary diffusion coefficient. Rotary diffusion is the process which will occur once an external force that aligned the molecules is removed and gradual redistribution of the orientation takes place.

In flow birefringence one can measure the total birefringence at different shear rates and/or the extinction angle χ. This extinction angle is obtained when the polarizer and analyzer are set at $90°$ and four dark areas are seen in the emerging light at the four corners of the cross of isocline. The extinction angle is the angle between the cross formed by the planes of polarization and the cross of isocline. At low shear rate this approaches $45°$ and at high shear rate (β) this should approach zero. The extinction angle is related to the rotary diffusion constant by

$$\chi = 45° - \frac{1}{12}, \quad \frac{\beta}{\theta} + \cdots \qquad (46)$$

for small β/θ values. For larger shear rates χ also depends on higher powers of β/θ. Hence a plot of χ vs β can yield a curve, the initial slope of which contains the rotary diffusion coefficient.

Varga and Gergely (*248*) studied the flow birefringence of hyaluronic acid. The birefringence was found to be positive with respect of the longitudinal axis of the molecules. The extinction angle decreased with shear rate and also with concentration. At sufficiently low concentration, however, where intermolecular interactions disappeared, the extinction angle became concentration independent. The rotary diffusion coefficient was linearly dependent on shear rate. The authors explained this as due to the polydispersity of their sample since for monodisperse rigid ellipsoids θ should be independent of β. The shorter molecules with higher rotary diffusion coefficients will orient only at higher shear rates.

The birefringence was increased with shear rate at low concentrations, but more than linearly. This was considered to indicate that the deformation of the molecules under the shear gradient was counteracted by the molecular interactions at higher concentrations. The length of the molecule at high and low shear rates were relatively close to 1750 and 1550 Å, respectively. This could have been interpreted in two ways: (1) If the sample was not very polydisperse, this would indicate a relatively high stiffness of the hyaluronate molecule, and (2) in the case of polydispersity, such results could have been obtained if the molecules were flexible random coils.

The authors considered the second case to be true mainly on the basis of the argument that hyaluronic acid, with a molecular weight of 10^5, should have a length of 3000 Å if highly extended. Therefore, they argue that a length of 1000 Å found by sedimentation–diffusion studies must represent a coiled state and the length of 1600 Å found by flow birefringence should represent the degree of uncoiling at the different shear rates.

Earlier investigators of the flow birefringence of hyaluronic acid considered the molecules extended with relatively high intrinsic stiffness (*249,250*).

With alginic acid the extinction angle of flow birefringence showed strong concentration dependence while the birefringence values ($\Delta n/C$) were concentration independent (*233*). In another experiment on alginic acid the flow birefringence was investigated in the presence and absence of added $CaCl_2$ (*251*). At low shear rates the extinction angle was lower in the presence of 0.1 M $CaCl_2$ than in its absence, but the situation reversed itself at higher shear rates. The intensity of birefringence was slightly lower in the presence of 0.1 M $CaCl_2$ than in its absence throughout the measured shear rates. The general trend of decrease in extinction angle and increase in the intensity of birefringence with increasing shear rate was observed in both experiments. In the presence of 0.2 M $CaCl_2$, however, the angle of extinction went through a maximum, after which stepwise decrease occurred with increasing shear rate.

The behavior of alginate in water and in 0.1 M $CaCl_2$ solution can be interpreted as those of elongated rigid molecules, the difference being that in 0.1 M $CaCl_2$ solution a sufficient degree of crosslinking occurred through Ca^{2+} bridges to alter the molecular weight distribution.

In the 0.2 M $CaCl_2$ solution, however, definite oriented structures are built up (similar to Thiele's ionotropic gels, see Section VII.B) which maintain their orientation over a certain range of shear rates. Similar experiments on pectinic acids were reported by Deuel et al. (*196*). The flow birefringence increased with shear rate and molecular weight. Aggregation of pectinic acid molecules by added electrolytes was also observed by a sudden increase in the birefringence and extinction angle.

VI. STRUCTURAL PROPERTIES IN THE SOLID STATE

Molecular parameters of single molecules of acidic polysaccharides obtained from measurements on dilute solutions have great instrinsic value in studying structure. However, these parameters definitely change when in more physiological conditions (concentrated solutions and gel state) the intermolecular interactions become important. Evaluation of the same parameters under these conditions becomes increasingly difficult due to the necessity of separating the variables.

At the other extreme, one can get information on the physical chemistry of the acidic polysaccharides in the solid state which will provide data on the intrinsic characteristics of molecular aggregates and of intermolecular interactions without having to deal with solute–solvent interactions.

One can then intrapolate from both extremes to describe physiological conditions. For one, the single molecular parameters can be considered together with the physical interpretations of the second and third virial coefficients of the expressions describing the experimental data on dilute solutions. For the other, the solid-state parameters can be followed up during solvent molecule sorption in the solid matrices and when the transition from solid-to-gel-to-solution occurs. For this reason, the physical chemistry of acidic polysaccharides in the solid state presented in this section is written with the idea that the same techniques discussed here can be used to follow up the swelling process and the eventual solvation process.

A. X-Ray Diffraction

If polymer molecules aggregate into crystallites or, as is the case with most proteins, into single crystals, X-ray diffraction is the primary tool to evaluate the structure. No single crystals of acidic polysaccharides have been prepared as yet, hence no detailed structural analysis has been reported for any of these compounds. However, a few of them form matrices in which a certain degree of crystallinity exists. These preparations can be studied by the Debye–Scherrer powder technique; the interplanar distances of the reflections can be indexed once a unit cell is assigned and elucidated. In most cases an oriented fiber pattern could be obtained from acidic polysaccharide preparations which, with the elucidation of the layer lines, indicates the repeating period of the fiber.

Thus the structure of the pectic acid (polygalacturonic acid) was found to have a threefold screw symmetry with a fiber axis identity period of 13.1 Å. This corresponds to a projection of the pyranose ring onto the chain axis of 4.37 Å, indicating that the carbon–glycosidic oxygen bond makes an angle of about 90° with the plane of the pyranose ring (*252*). Alginic acid has a twofold screw symmetry, an identity period of 8.7 Å, and the same projection of the pyranose ring into the chain axis as the pectin acid (*253*).

Hyaluronic acid has a fiber identity period of 11.98 Å which corresponds to the projection of the repeating unit, hyalobiuronic acid, onto the chain axis. This indicates that the 1-3-β linkage in the hyaluronic acid is a very stiff one, the glycosidic oxygen being almost in the plane of the pyranose ring. Compared to this, the rigid cellulose molecule and its derivatives have an identity period of 10.3 Å for two glucose units in the 1-4-β linkage and a 20° angle between the glycosidic oxygen–carbon bond and the plane of the pyranose ring (*254*).

Among the chondroitin sulfates investigated, only a fraction of chondroitin-4-sulfate isolated from bovine trachea showed crystallinity, and this fraction had a sulfate:chondroitin ratio of 1:3 (*24*). X-ray diffraction data showed that the repeating period is 9.8 Å, corresponding to the disaccharide units. Strong $hk3$ reflections indicated that the polymer is stereospecifically sulfated on every third chondroitin moieties, and the chains are packed in the crystallites in such a fashion that the calcium sulfate groups in each chain are shifted by one chondroitin moiety relative to the same moieties in the neighboring chains (*25*).

X-ray diffraction on 4-*O*-methyl-D-glucuronoxylanes from different sources yielded identical fiber diagrams. These polymers contain the glucuronic acid as a side chain on about every sixth xylose unit on the C_2 position from white birch and 30–40% less glucuronic acid in xylans from white elm and milkweed floss. Marchessault and Liang (*255*) concluded, therefore, that only the xylane backbone residues take part in crystallization. From the layer line spacing, they calculated a repeating unit of 15 Å composed of three xylose residues having a threefold screw symmetry. The unit cell had a hexagonal symmetry.

X-ray diffraction studies on whole carrageenan and on the λ and κ forms yielded a fiber repeat period of 25.2 Å. This probably corresponds in the λ form to three disaccharide units since strong 8.4 Å meridional reflections were also obtained (disaccharide length). The reason for the long repeating period is probably the variation of the number of sulfate groups on the galactose residue. The κ form is branched and the fiber repeat period represents two trisaccharidic units, each composed of two sulfated galactose units in 1-3-α linkage and one 3,6-anhydro-D-galactose residue linked 1-4-β linkage. Within each 25.2 Å period a single 3,6-anhydro-D-galactose residue appears to be a side chain attached to C-6 of the sulfated galactose moiety. The unit cell of λ carrageenan contained one chain, that of κ four chains, while that of the whole carrageenan contained two κ and two λ chains (*256*).

Other pertinent data obtainable from X-ray diffraction patterns are less amenable to accurate interpretation. One can obviously obtain the symmetry and the dimensions of the crystal unit cell. One also knows the density of the solid. If the polymer was 100% crystalline, these data would allow us to calculate the number of chain segments in each unit cell, etc. However, all these polymers have only a certain degree of crystallinity and, therefore, the calculations of the number of chain segments per unit cell are at best educated guesses and so are the interchain distances. Sodium pectate has a hexagonal symmetry with $a = 16.2$ Å and $c = 13.1$ Å, volume 2980 Å^3, and 12 galacturonide residues per unit cell (*252*). Sodium hyaluronate also has a hexagonal unit cell with $a = 12.66$ Å, $c = 11.98$ Å, volume 1662 Å^3, and 4 hyalobiuronic acid residues per unit cell (*254*). The crystalline chondroitin-4-sulfate fraction has orthorhombic symmetry with dimensions of $a = 14.06$, $b = 8.62$, $c = 9.80$ Å, and volume 1188 Å^3 with 3 chondroitin chain segments per unit cell (*25*).

In general, one can state that acidic polysaccharides from plant sources tend to yield better crystal structures than those from animal sources mainly because the latter are intimately associated with proteins which may disrupt the three-dimensional ordered arrays. Bulky side chains are a definite hindrance to crystalline organization. Sulfate groups in chondroitin sulfates, keratosulfate, heparin, etc. have to be considered as steric hindrances to crystalline alignments.

B. Absorption Spectra and Dichroism

Structural characterization of acidic polysaccharides by their absorption spectra is usually performed in solution as well as in the solid state. It is well known that the environment will cause frequency shifts of absorption bands. The main reason why this topic is dealt with in this section is that in contrast to small molecules for which assignments of the absorption bands are made primarily on a comparative basis with other derivatives or on single crystals, the polymeric matrices in the solid state allow certain orientations to be made by strain or other techniques and hence polarized absorption spectra (dichroism) is a principle help in absorption band assignments.

Although most of what follows is applicable to absorption spectra in the whole range of electromagnetic radiation, i.e., ultraviolet (electronic transitions) to microwave region (rotational transitions) in the case of acidic polysaccharides, the preponderance of experimental data is concerned with vibrational transitions in the infrared region. Hence only this region will be dealt with extensively.

One would assume that in order to evaluate a complex infrared spectrum of a polymer, one would have to know the detailed assignment of absorption bands of the monomeric units. This is probably true for the "complete" assignments and, therefore, works on simple sugar moieties such as those of Tipson and Isbell are of great help (257,258). However, some vibrations of simple crystalline sugars are "damped out" in the polymeric spectrum as a consequence of imperfect crystallinity (259). The result is that the absorption bands of polymeric material are usually quite broad (10–20 cm^{-1}), and high resolving power spectrometers are necessary to resolve the amorphous and crystalline contributions of the same vibrational mode (260). In addition to the necessity of comparing spectra of derivative acidic polysaccharides and their deuterated compounds to those of simple sugars, the dichroic ratio of the oriented polymeric network plays an important part in the assignment of bands. By dichroic ratio we mean the ratio of absorption intensities of a band of oriented polymers when the infrared radiation is polarized parallel and perpendicular to the orientation of the sample

$$D = a_\perp/a_\parallel \qquad (47)$$

The absorption at a characteristic wavelength is proportional to the interaction between the dipole change vector, $\bar{M}$, of the specific group vibration and the electric vector, $\bar{P}$, of the electromagnetic vibration. Thus, when $\bar{M}$ and $\bar{P}$ are parallel, a maximum absorption occurs; when they are perpendicular, no absorption should occur. Hence, the dichroic ratio indicates the orientation of a specific group with respect to the orientation of the sample film or fiber.

If one can resolve amorphous and crystalline contributions to an infrared band and obtain crystal orientation from X-ray data as supporting evidence, one can evaluate the amorphous orientations as well. Theoretical treatment of such phenomena has been worked out (261,262).

Mainly on the strength of works by Marchessault and Liang on cellulose, xylans, chitins, and similar polysaccharides with polarized infrared radiation, we can assign the absorption bands of the acidic polysaccharide backbone with relative ease (263–266). These are:

OH out-of-plane bending	650–685
NH out-of-plane bending	730
CH_3 wagging along chain	975
C—O stretching	980
Antisymmetrical in phase ring stretching	1110–1130
Antisymmetrical bridge C—O—C stretching	1155–1165
OH in-plane bending	1205–1215
CH_2 wagging	
CH bending	
Symmetrical CH_3 deformation	1280–1320
Amide III band	
OH in-plane bending	1317–1336
COO^- salt symmetric stretching	1425
CH_2 symmetric bending	1420–1460
Amide II band	1555
COO^- salt antisymmetric stretching	1600
Amide I band	1652
COOH stretching	1736
CH_2 symmetric stretching	2840–1850
CH stretching	2873–2914
CH_2 antisymmetric stretching	2929–2935
CH_3 stretching	2962
NH stretching	3264
OH stretching	3430–3480

With regard to the position of the sulfate groups, an extensive amount of work has been done on sulfate esters of both monosaccharides and polysaccharides. Absorption bands around 1240 cm^{-1} are attributable to modes of vibration (symmetric and antisymmetric stretching) involving S—O linkage while those in the 800–900 cm^{-1} involve C—O—S linkages (267). The latter region, the so-called "finger print" region, is especially important because sulfate groups at different carbon atoms of mucopolysaccharides have either equatorial or axial configuration. Hence, the correct assignment of bands in this region is useful for structural analysis. Chondroitin-6-sulfate (C) has absorption bands at 1000, 820, and 775 cm^{-1} while chondroitin-4-sulfate (A) and dermatan sulfate (B) have bands at 928, 852, 725, and 928, 855, and 719 cm^{-1}, respectively. The lower frequencies within the 800–900 cm^{-1} region are attributable to equatorial configuration and the higher ones to axial configuration. Hence chondroitin-4-sulfate and dermatan sulfate have axial sulfate groups on carbon 4. Chondroitin-6-sulfate has its sulfate in the equatorial configuration (268–270).

Infrared dichroism studies on sodium hyaluronate revealed that the C—O—C bridges orient parallel to the direction of elongation rather rapidly while the CH$_3$ group orients itself perpendicularly somewhat more slowly. The indication is, therefore, that the hyaluronate chains are rather stiff and orient themselves by turning in the direction of stretch rather than by uncoiling. The bands at 1420 and 1655 cm^{-1} representing the ionized carboxyl and amide I, respectively, showed only slight parallel dichroism, indicating that their dipole change vector is at an angle of about 30° to the direction of stretch (271). Similar results were obtained in xylans where the C—O—C bridges have shown strong parallel dichroism but the carboxyl group no preferential polarization. Hence, in the structural assignments the latter was assumed to have a dipole change vector at 45° to the backbone of xylan (264).

Since in the spectroscopy of polymers, because of their complex nature, one does not have an excess of data to help in the elucidation of structure, the paucity of supporting Raman spectra is a definite disadvantage. It is quite difficult to obtain Raman spectra on high polymers. However, recently some technical advances have been made and, therefore, in the near future one can expect research interest in this direction (272).

Similar considerations apply to the ultraviolet spectra of acidic polysaccharides. One would expect both the carbonyl and sulfate groups to contribute, and one can predict that the $n - \pi^*$ ~2800 Å and $\pi \rightarrow \pi^*$ ~ 1870 Å transitions of the carbonyl group can be identified in a manner similar to that for synthetic polymers such as polyvinyl acetate (273). Another unexplored area of molecular spectroscopy of acidic polysaccharides is in the microwave region. As with synthetic polymers, this region is capable of yielding information on certain segmental vibrations of the backbone or hindered rotations of side chains. Again, the expense of instrumentation provides a barrier to structural research of acidic polysaccharides at these frequencies (274).

C. Strain Birefringence

Strain birefringence is the result of optical anisotropy in a solid form which has been oriented, usually by elongation. It represents the difference in refractive indices parallel and perpendicular to the axis of orientation, and it is measured with monochromatic light by placing the oriented film between a polarizer and a compensator (half wave plate or Babinet compensator) and obtaining the retardation or the number of waves of path difference. The retardation, N, is related to birefringence by

$$N = (n_\parallel - n_\perp)d/\lambda_0 \tag{48}$$

where $(n_\parallel - n_\perp)$ is the birefringence, d the thickness of sample, and λ the wavelength of the light. In the solid state or gel this is not a parameter of a single molecule as in flow birefringence, but the property of molecular aggregates. Hence, there are a number of contributions to the total birefringence: (1) intrinsic-crystalline, (2) intrinsic-amorphous, (3) form, and (4) adsorption birefringence. The intrinsic crystalline birefringence is obviously present only if the polymer matrix contains short range order and can be separated in conjunction with X-ray diffraction studies and knowledge of the refractive indices along the axes of the unit cell (275). Both the intrinsic crystalline and amorphous birefringence are due to the orientation of the crystallites and the amorphous chains, respectively, relative to the axis of the macroscopic orientation. Form birefringence is the result of anisotropic geometrical shapes (rodlets, platelets) embedded in isotropic medium, and it can be evaluated by imbibing liquids of different refractive indices in the polymer matrix to yield Wiener curves (parabolic curves when the birefringence is plotted versus the refractive index of the medium) (276) (Fig. 14). At the minima or maxima of the Wiener curves only intrinsic birefringence occurs, and the refractive index of the imbibed liquid at this minima corresponds to the refractive index of the anisotropic body perpendicular to its long axis. Adsorption birefringence occurs when the imbibing liquid itself has anisotropic polarizabilities and is strongly adsorbed on the surface of the molecules or molecular aggregates. This can be avoided by careful selection of the imbibing liquid.

From the point of view of molecular structure, intrinsic or orientation birefringence is the important one. But because of the necessity of separating this from form birefringence, the technique is not strictly in the solid state but represents the transition from solid to gel state which is the topic of the next section. In spite of that, we are discussing it here because most of the measurements were made on solid films (imbib-

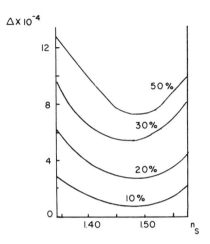

Fig. 14. Wiener curves of form birefringence of sodium hyaluronate. The total birefringence, Δ, is plotted against the refractive index of the imbibing solvent at different percent elongations of the hyaluronate films.

ing medium air) and no attempts were made to separate the intrinsic and form birefringence.

The acidic polysaccharides investigated to the greatest extent were the pectic and alginic acids. Negative birefringence is found for pectic acid [i.e., the refractive index along the polymer backbone is smaller than that perpendicular to it (*277–279*)]. Alginic acid, on the other hand, yields positive birefringence (*279–281*). Sterling has evaluated the intrinsic birefringence of both by obtaining Wiener-type curves with imbibed liquids of different refractive indices (*282*). An explanation of the difference between the behavior of pectic and alginic acid was found by Sterling considering the configuration of the repeating units. In pectic acid the C—O bond polarizabilities at C_2 and C_3 and the C—C polarizabilities at C_5—C_6 are axial, while in alginic acid only the C—O at C_2 is axial and the other two are equatorial. Sodium hyaluronate shows strong positive birefringence (*283,284*). On the other hand, chondroitin-4-sulfate has very little birefringence. Comparing this to the behavior of hyaluronic acid, one concludes that the sulfate group must add strongly to the polarizability perpendicular to the polymer backbone (*285*).

DerSarkissian and Bettelheim (*284*) evaluated the intrinsic and form birefringence of sodium hyaluronate at different elongations. The intrinsic birefringence up to 25% elongation showed an orientation process resembling a rubberlike material, indicating that at first the nonhydrogen bonded parts of the amorphous chains orient in the direction of stretch in a rubberlike behavior.   Above that extension the intrinsic birefringence slowly approaches a maximum.   On the other hand, the form birefringence is increasing, also indicating that to a large degree the anisotropic platelets (hydrogen-bonded aggregates) turn directly with their long axis toward the direction of stretch.

VII. BEHAVIOR OF GEL AND CONCENTRATED SOLUTIONS

Acidic polysaccharides in biological conditions often appear in the form of gel or concentrated solutions.   Although these two have different physical properties (the gel being in essence a solid phase), under the above heading we are referring to states where the polymer concentration may vary from 1–50% by weight.   The main difference between the two phases will be in their flow properties (viscosity).   These biological states can be approached from two extremes: (1) from the solid state by water vapor sorption and subsequent swelling, and (2) from the dilute solutions by concentration, coacervation, and flocculation.

One could define the phase transition between concentrated solution and gel from a phenomenological point of view as a transition where both components (solute and solvent) become continuous.   The polymer continuum may be established by random entanglement of the primary chains, by certain chemical crosslinking, or by the establishment of regions with short-range order such as microcrystallites.   In any case the continuous polymer network will be held together by multiple types of forces.   In acidic polysaccharides these forces range from ionic and hydrogen bonds to ion–dipole and dipole–dipole interactions.

A. Swelling by Vapor Sorption

Probably the simplest way to study these gels and concentrated solutions is to follow their transitions from the solid . state.   With acidic polysaccharides this is most successfully done by water vapor sorption on the polymer matrix.

Sorption isotherms of water on acidic polysaccharides yield B.E.T. (Brunauer, Emmett, and Teller) II type multilayer curves (Fig. 15) except in cases such as pectic acid where the high degree of crystallinity provides strong cohesive forces and the shape of the isotherm is B.E.T. III curving smoothly upward (286).

The amount of water vapor absorbed at a certain relative humidity depends both on the chemical structure (i.e., number of polar groups available on the polymer) and the physical state of aggregation.   Highly crystalline sodium hyaluronate (254) had less sorptive capacity than

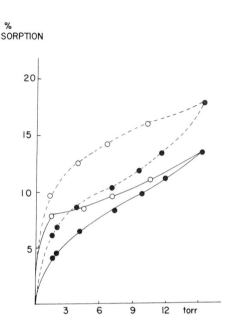

Fig. 15. Water vapor sorption isotherms of potassium hyaluronate (——) and sodium hyaluronate (- - -) isolated from umbilical cords. Sorption isotherms (●); desorption isotherms (○).

amorphous sodium hyaluronate from umbilical cord (286). The potassium salt of hyaluronic acid from umbilical cord had less sorptive and retentive capacity than the sodium salt (286). For instance, highly esterified N.F. pectin absorbed less water than deesterified sodium pectate, and pectic acids sorbed even less at low water vapor pressures (286a). Sodium alginates dried by ethanol dehydration had a greater moisture sorptive capacity than those dried by heating. The water vapor sorption capacity increased with the degree of polymerization and decreased with the degree of acidity. A maximum moisture content of 36% was found at 80% relative humidity (287). The sorptive capacities of sulfated mucopolysaccharides at low vapor pressures following order: heparin > chondroitin-6-sulfate > dermatan sulfate > chondroitin-4-sulfate. At higher vapor pressures the order between chondroitin-6-sulfate and heparin was reversed. In all these mucopolysaccharides the H_2O sorptive capacity was higher than that for D_2O (288).

Since B.E.T. type isotherms are obtained the "monolayer" can be evaluated. For pestic substances there is 1 mole of water/mole of monosaccharide (289,290). For mucopolysaccharides the values are 1.2 for chondroitin-4-sulfate, 1.9 for dermatan sulfate, 2.2 for chondroitin-6-sulfate and 2.4 for heparin, each in units of moles of water per disaccharide repeating, unit (*288*).

Most sorption isotherms show a strong hysteresis indicative of the nonequilibrium nature of the sorption process, which may be the result of structural differences in the matrix when a certain swelling stage is approached from the solid state or from a more dilute gel state. However, these non-equilibrium processes are usually stationary; i.e., when approached from the same reference state, they are reproducible. The hysteresis loop may reflect the mechanical constraints introduced in the swelling gel (*291*). From this point of view it is interesting to observe the relationship between the sorptive capacity and degree of hysteresis of the mucopolysaccharides. The width of the hysteresis loop follows exactly the opposite order of the sorptive capacity; (chondroitin-4-sulfate > dermatan sulfate > heparin > chondroitin-6-sulfate). This may be interpreted in the following manner: In a tight polymer matrix, such as chondroitin-4-sulfate, the smaller amount of water vapor sorbed may be more constrained by the polymer network than the larger amount sorbed in a loose network, such as chondroitin-6-sulfate.

In spite of the irreversible nature of the water vapor sorption in acidic polysaccharides and because they are stationary nonequilibrium systems, one can calculate the differential enthalpy and entropy of sorption from the Clausius–Clapeyron equation as a first approximation. Such calculations assume that no entropy production occurs in the system, which is clearly a contradiction to the irreversibility exhibited by the hysteresis. Therefore, heats of sorptions calculated as isosteric heats are clearly overestimates and the calorimetric determinations yield lower values (*247*). However, the entropy production is a minimum because of the stationary nature of the nonequilibrium system, and due to this circumstance the approximation can be made that the maxima of the thermodynamic functions from isosteric heats and calorimetric measurements will be at the same water uptakes (*292*). The absolute values obtained at different coverages are not important because of the nature of the approximations and because in a swelling system both the configurational change of polymer and water will contribute to the entropy. In most cases the differential thermodynamic functions show more than one maximum with respect to coverage. The maxima of the thermodynamic functions would be indicative of completion of coverage of energetically different sorption sites. This can be interpreted in two ways. For instance, in pectic substances the first maximum appears where the B.E.T. equation indicated a monolayer coverage. A second and a third maxima could be interpreted as previously hydrogen-bonded sites opened up by swelling and made available for further sorption.   Highly esterified pectinate and xylan showed one maximum, sodium and calcium pectates two maxima, and pectic acid three (*188,289,290*) maxima.

On the other hand, in the case of mucopolysaccharides where all differential entropy functions showed two maxima, this could be interpreted as either indicating the presence of two energetically different sites on which the sorption occurs in two steps or that the water is sorbed in the monolayer by either one or two hydrogen bonds.   The latter is the preferred explanation on the basis of infrared and microwave spectroscopy of the sorption process (*293,294*).   With the sulfated mucopolysaccharides the second maxima appeared where the B.E.T. equations indicated the completion of monolayers.   In the case of a high molcular weight (2×10^6) hyaluronic acid, the sorptive capacities were on the order of that of chondroitin-6-sulfate, and the B.E.T. monolayer gave about 1 mole of water/disaccharide repeating unit with the differential thermodynamic functions yielding a first maximum at the B.E.T. monolayer and a second maximum at 3 moles/disaccharide repeating unit. Hence this polymer behaved more like the pectic substances.   However, after repeated sorption and desorption experiments the reproducibility of the isotherms was not good, indicating that certain changes occurred in the polymer matrix, possibly a degradation process (*295*).

In addition to the sorption isotherms and the data derived from them (monolayer coverage, differential thermodynamic functions, hysteresis), one can obtain additional insight about the solid–gel transition by following some physical chemical parameters during the sorption process.

In the case of crystalline solid the change in the X-ray diffraction pattern gives some insight on the swelling process of crystallites.   With sodium pectate, Palmer and his coworkers found changes in the equatorial spacings of the X-ray diffraction pattern up to 24% moisture content after which no further changes occurred.   In highly esterified pectins no change occurred throughout the whole range, while in pectinic and pectic acid changes occurred over the whole sorption process (*296,297*).   In our reinvestigation we found that equatorial spacing changed throughout the sorption process with all pectic substances (*286*).   The interpretation of this is that interchain spacings increased with water content and that this change usually followed the shape of the sorption isotherm.   This change, however, was the smallest with calcium pectate, indicating that the divalent cation acts as an interchain bridge restricting swelling (*290*). Meridional spacings also changed in sodium pectate and highly esterified sodium pectinate, thus showing an increase in the repeating unit length, but there was very little change in pectic acid or calcium pectate.   In the case of crystalline sodium hyaluronate, changes in the equatorial spacings only were observed (295).   Usually in each case the X-ray diffraction powder pattern sharpened with water content at low vapor pressures, but the sharp pattern disappeared close to saturation vapor pressures. This indicates that water adsorbed in the monolayer helps the crystalline organization, but when multilayer sorption occurred and increased the chain separation, an amorphous structure set in which finally results in solvation of the acidic polysaccharide chains.   Another physical parameter change during sorption is in birefringence.   This has been dealt with in the previous section on solids because the evaluation of form birefringence was necessary to estimate the intrinsic birefringence.

The question of the role of the different polar groups of acidic polysaccharides in the sorption process can be evaluated from infrared data. Water molecules sorbed on specific polar groups will perturb the vibrational modes and result in shifts of the absorption band to lower wave numbers and in enhanced integrated intensities.   Such investigation on mucopolysaccharides (293) indicated that all polar groups sorb water throughout the whole sorption process, usually in proportion of their abundance (i.e., OH groups most the COO^- and SO_3^- least).   Hence, there are no preferential sites.   However, a comparison between the sulfated mucopolysaccharides at equal coverage indicated that the carboxyl group is the most important binding site in chondroitin-6-sulfate and dermatan sulfate and less important in chondroitin-4-sulfate and heparin, while the opposite is true for the sulfate group at low and medium coverages.   Such data gives information about the nature of the solid gel (i.e., the less important a group is in water sorption, the more important it is in the polymer–polymer interaction in the gel).   A further parameter aiding the evaluation of gel network is the study of the change in the dielectric dispersion during sorption process.   It has been found that with polygalacturonic acid and chondroitin-4-sulfate the polymer matrix shows an increasing dielectric loss with increasing water content around 100 kc.   The relaxation times calculated from such studies indicate that the increase in dielectric loss is due to the enhanced backbone segmental motion of the polymer with increasing degree of swelling (298).

Besides vapor sorption studies, direct swelling studies have been performed on acidic polysaccharides in different solutions.

Laurent, Hellsing, and Gelotte (299) found that when hyaluronic acid was treated with 0.2% sodium hydroxide and swelled in a refrigerator overnight, the concentrations of acid being varied from 0.9 to 17.5 g/100 ml, gels were formed from all concentrations except the lowest.   The gels were all found to have high swelling capacities with distilled water.   The swelling was dependent on ionic strength. The absorption of water decreased greatly with increasing sodium chloride concentration.

Miyamoto and Sasaki (*300*) investigated the degree of swelling of alginic acid powder in various electrolyte solutions. The results showed that the degree of swelling increased with the cation present in the order sodium > potassium > ammonium, and the anion present in the order nitrate > iodide > isocyanate > bromide > chloride.

B. Ionotropic Gels

A second approach to gel structures of acidic polysaccharides can be followed through the studies of Thiele and his coworkers. Ionotropic gels were obtained through polyvalent cation diffusion into an acidic polysaccharide solution. These ionotropic gels have shown definite organization, and when the diffusion was governed by centrifugal forces the gel had an inner anisotropic section and an outer isotropic front zone which had high swelling capacities. Alginic acid, pectic acid, carboxymethyl cellulose and carboxy cellulose all build ionotropic gels reversibly from solutions with divalent cations (*301,302*). Most of the work has been done on alginic acid, and the specific binding of the counterions have been followed by study of equilibrium diffusion against $NaNO_3$. The reversible destruction of the ionotropic gel and the diffusion capacity of the binding counterions followed the order Pb > Cu > Cd > Ca > Zn > Mg (*303*). The studies indicated that the counterions interact with the acidic polysaccharides in four different manners which are in equilibrium with each other: (1) strong specific homopolar binding, (2) a heteropolar binding of the screening counterions, (3) and electrostatic binding of the smeared-out counterion cloud, and (4) free ions in solution.

Usually three different zones can be seen in the building of ionotropic gels: (1) the solution, (2) the frontal zone, and (3) the ionotropic gel. Only the ionotropic gel is birefringent although upon addition of acid the frontal zone also becomes birefringent while the gel increases its birefringence. The frontal zones are isotropic gels which do not show either thixotropy or thermal reversibility. As the crosslinking through divalent ions proceeds the swelling capacity decreases in the frontal zone (*304*). The mechanism of the formation of ionotropic gels was studied through birefringence and turbidity measurements. The total birefringence is a composite parameter of five contribution: (1) orientation birefringence of the polymer chains, (2) form birefringence of the crystallites embedded in the medium, (3) intrinsic birefringence of the crystallites, (4) hydration birefringence, and (5) degree of crystallinity. Thiele and Schacht have exchanged the counterions for H^+ ion and, observing the birefringence change in this process, they assumed that contributions (2), (3), and (4) were kept constant, and thus they evaluated the orientation and crystallinity contributions during the ionotropic gel formation (305).

They concluded that the mechanism of orientation occurs at the sol-frontal zone boundary and, through a membrane potential, the crystallites formed are oriented perpendicular to the direction of ion diffusion. Under certain concentration conditions of acidic polysaccharides and counterions, capillary structures could also be obtained in ionotropic gels which were then investigated in a polarizing microscope (306,307).

Bettelheim has studied the kinetics and the structures of ionotropic gels of polygalacturonic acid and Ca^{2+} ions with light scattering (308). At the boundary of sol–gel (i.e., at the incipient gel formation boundary) highly organized structures appeared which gave diffraction patterns with monochromatic light (5461 Å). The diffraction patterns were interpreted as coming from capillaries which are rather uniform in diameter ($12\ \mu$) and running perpendicular to the direction of diffusion. The length of the capillaries extended to a thousand microns. Their formation was explained by the mechanism described below.

The diffusion of Ca^{2+} ions into the polygalacturonic sols align the polyelectrolytes into crystalline regions through Ca^{2+} bridges with the consequent dehydration process. The water molecules coalesce and form syneretic droplets, and the droplets coalesce and form channels. The aggregated polygalacturonic acid domains surround these water channels, most probably in a helical conformation since the high birefringence of the polyelectrolyte molecules disappear in the aggregate formation. These perfectly aligned channels and their dimensions were also confirmed by microscopic observations.

Further crosslinking of the gel by the still diffusing Ca^{2+} ions make the highly organized structures disaligned so that beyond the incipient gel boundary the light-scattering patterns do not resemble diffraction patterns anymore, although their anisotropy indicates that the disalignment of channels is not completely random.

In general five different structures have been obtained by Thiele et al. (309) in ionotropic gels: (1) solution, (2) striation, (3) lenses, (4) capillaries, and (5) membranes. The similarity between such ionotropic gels and biological structures such as eye lenses, cornea, aorta, skin, tooth, and hair was pointed out. In experiments the acidic polysaccharides of these tissues were dissolved in different solvents and through divalent ion diffusion rebuilt from the acidic polysaccharide sols (310). It is, however, doubtful whether the rebuilt ionotropic gels represent the original gel structures in the different parts of tissues. It has been shown that the complex gel structures in tissues are more influenced by the collageneous fibers and fibril aggregates than by the acidic polysaccharide matrix. This is especially the case in dilute gels such as the vitreous humor of the eye where the only structure demonstrated is the concentric layer around the Cloquet canal largely associated with collagen aggregates (*311*). In more condensed gel systems, such as the cornea, the orientation of the collagen fibrils in the lamellae is again the structure-determining factor (*312*). In bovine lenses, layer-like structures give rise to a light-diffraction phenomenon (*312a*).

C. Gelation by Condensation

A third way of achieving gel structure is to start with dilute solutions and enhance the polymer chain entanglement by (1) cooling, (2) evaporating the solvent, (3) adding dehydrating agents, and (4) the combination of these. These techniques are used in achieving most of the commercially important gels such as agar-agar, pectin, and alginates.

Gels may be classified as irreversible and reversible. Irreversible gels are stable towards heat and additional solvent. The polymer molecules are crosslinked by covalent bonds. Reversible gels are dispersible upon heating and usually are soluble in excess solvent. Here crosslinks are secondary valence or ionic bonds.

Linear chain molecules generally show much better gelation properties than branched ones. Uncoiled molecules react preferentially with each other whereas coiled molecules prefer to react with different segments of the same molecule. Coiling will be influenced by the charges on the macroions and side groups of the chain molecules. Side groups or branching may hinder contact of two molecules for steric reasons and may prevent crosslinking. The main reactive groups of acidic polysaccharides are hydroxyl, carboxyl, and sulfate and methyl ester groups which can participate in reaction with bifunctional reagents and which will produce covalent bridges of highest stability between the polysaccharide molecules forming a covalent three-dimensional gel network. The reaction is governed by the concentration of the gelling agent and is generally irreversible. The reversible gels probably have direct chain-to-chain contact involving coulombic interactions between ionic groups over large distances or hydrogen bonds over small distances. In the reversible agar-agar gels the combination of repulsive coulombic forces and attractive hydrogen bonds maintain the structure. This has been vividly demonstrated by nuclear magnetic resonance studies. The water protons of agar-agar gels gave a broader signal with decreased amplitude relative to the water standard.

The line widths varied from about 5 Hz for 1% to 50 Hz for 10% gel. A decrease in the area of the curves of the agar-agar compared to the same percentage of water in D_2O indicated that some of the hydrogens of water in the gel are bonded. Line widths of the agar gels broaden as the agar cools (313). Balazs et al. have studied hyaluronic acid gels and reported no broadening of the water signal (314). Therefore, the line broadening of agar cannot be explained in terms of any single gelation or viscosity effect. Agar gels (3%) prepared in $1 M$ $(CH_3)_4NCl$ showed the same broadening of the water peak but the methyl peak was not significantly broadened. This ruled out the possibility of water enclosed in small heterogeneous compartments. The line broadening was due to the decreased mobility of water protons in agar gels relative to pure water, implying some intermediate water structure where water molecules are hydrogen bonded to hydroxyl groups of the acidic polysaccharide chain.

The same conclusion was reached by Arakawa (315). From stress relaxation data on agar-agar gels at different temperature, he obtained the activation energy. He found it was substantially the same as the value of H bond in liquid water.

The gelation of alginic acid by other than crosslinking with multivalent cations (ionotropic gels) was studied at concentrations sufficient to produce a maximum viscosity above 10,000 cP. The viscosity maximum occurred at pH's lower than necessary for 100% neutralization of the —COOH groups. Urea and NaCl eliminated gelation. The heat of formation for gelation (crosslinks) was 8 kcal/mole. A balance between forces of H bonding attraction and electrostatic repulsion accounts for the gelation behavior (316).

Gels of hyaluronic acids were formed when solutions of different concentrations at different pH's were subjected to a gravitational field in an ultracentrifuge. The sedimented hyaluronic acid formed a viscoelastic putty which showed "healing" properties; i.e., when pieces were cut off from the putty they could be reattached by pressing the pieces together.

A maximum was found at pH 2.5 in the storage modulus, G', and loss modulus, G''. The storage modulus measures the energy stored in the material in the form of elasticity, and the loss modulus measures the energy dissipated in the form of heat as viscosity.

Only one relaxation machanism was found to be operative in the hyaluronic acid gel and that was the breakdown of the highly elastic hydrogen-bonded network followed by viscous flow. Hyaluronic acid smaller than 10^5 molecular weight did not yield viscoelastic bodies. Similar viscoelastic gel is found in fish eyes. The gel was composed of ichthyosan, an acidic polysaccharide quite similar to hyaluronic acid, and it acts as a damping body on which the lens of the fish eye floats (317).

The most commercially important gels of pectic substances are the subject matters of numerous articles in the food industry (69). These are usually formed in acidic media, pH 3–4, and in high concentrations of simple sugars. The sugars act as competing agents for the solvent (water) and thereby decrease the solvation of the polygalacturonic chain; the acid medium, on the other hand, decreases the ionization of the carboxyl groups and thereby enhances the hydrogen bond formation between the polymer chains. Acid and methyl ester groups are not the only essential elements for the gel formation of pectins; the configuration of the secondary hydroxyl groups at carbon atoms 2 and 3 is also of importance (318). Because these hydroxyl groups are in a *trans* configuration, the formation of intermolecular hydrogen bridges is favored. Alginic acid showed less tendency to form gels, probably because the neighboring hydroxyl groups are in the *cis* position. The experiments indicated that not points but zones of attachment between the chain molecules are necessary for gel formation. This would explain the fact that a few acetyl groups can interfere with the formation of crystalline regions. A more extensive substitution with acetyl groups produced a more regular structure and again made gel formation possible. As little as 2.5% acetyl, representing one acetate group for about eight galacturonide units, was sufficient to eliminate the gelling ability of peotin. This indicated that a certain degree of regularity along the pectin chain is necessary. If the chain is too regular, as in freely methylated pectin, the attractive forces along the chain are sufficient to cause association and precipitation of the pectin (319).

Completely methoxylated pectins form gels with sugar alone, with 70% methoxylation and higher they form gels with additions of acid and sugar at a pH of 3 to 3.4 (69). At lower degrees of esterification more acid is required, and below 50% esterification pectins do not form gels with sugar and acid. These pectinic acids form gels in the presence of calcium ions and other polyvalent cations without sugar and acid by formation of ionic bonds.

Sugar appears to serve primarily as a desolvating agent since 99% of the sugar in a gel can be extracted with ethanol without a change in the gel dimensions (320). Harvey (321) reports that a concentrated sugar solution acts as an ionizing solvent to almost exactly the same extent as an equal volume of water. It has been shown that sugar reduces the acidic hydrolysis of agar-agar (322). The amount of Ca ion shows an optimum for pectins with a methoxyl content from 5.5 to 6.5% (323).

The dependence of the gel stability on the Ca ion concentration was illustrated when no change occurred to a calcium pectinate gel immersed in fifteen volumes of distilled water. However, when sodium oxalate was added to the water the Ca ions linking the pectinic acid molecules were removed by precipitation as the oxalate diffused into the gel and the gel dissolved. Evidently the Ca ion is bound sufficiently strongly to the car-boxyl groups so that it cannot diffuse out of the gel even in excess solvent.

The size of the hydrated cation in gelation may be an important factor as evidenced by an X-ray study on carrageenan (256). The diameter of the hydrated cation must be smaller than 4 Å, consequently Na and Li ions will not produce gelation. Potassium can fit between the sulfate groups and can cause association of the molecules by neutralization.

D. Complex Sols, Gels, and Flocculates

In the above discussions we have proceeded from the simplest unicom-plex colloidal system (swollen solids) to dicomplex system of colloid anion and microcation (ionotropic gels) in the terminology of Bungenberg de Jong (324). The metachromatic dye–polyanion complexes are also di-complex colloidal systems of acidic polysaccharides (see also Section IV.C). In this case the dye molecules are in orderly alignment on the polyanion surface. This results in a shift in the wavelength of the trans-mitted light from the ortho-chromatic level of a longer wavelength. Meta-chromasia is used to a great extent in microscopy and histochemistry of tissues containing acidic polysaccharides. The physical state of the dye–polyanion complexes (sol, gel, or flocculates) is determined largely by concentration and environmental conditions. The anionic sites and the bound cationic dye molecules are in stoichiometric relation (325). The interaction is primarily electrostatic, and in the case of the acidic polysaccharides the dye binding capacity has the following order: $R—CH_2COO^- < R—COO^- < R—OPO_3^{-2} < R—OSO_3^-$. However, since most of these dyes (for example, Azure A) have large hydrophobic por-tions the assumption is that mono and multilayer sorption on the poly-anion surface occurs with the hydrophobic part of the dye molecules pro-viding lateral interactions among themselves accounting for the "red shift." The value of these lateral interactions are of the order of about 8 kcal/ mole dye (326). Similar interaction between quaternary ammonium compounds which have curare-type physiological actions and acidic poly-saccharides have been reported (327–329).

More important interactions between quaternary ammonium detergents and polyanions have been utilized to separate and fractionate acidic poly-saccharides from tissues. Mucopolysaccharides can be precipitated quan-titatively from solutions by cationic detergents such as cetylpyridinium chloride (330). These complexes are soluble in salt solutions above a certain critical salt concentration. This critical salt concentration depends on the nature of salt, the structure and molecular weight of the polyanion, and the chain length and cationic group of the detergent (*331–333*). In the separation of acidic polysaccharides a column of porous material (cellulose) is saturated with the quaternary ammonium compound. The aqueous polysaccharide solution is added and washed with the detergent solution. The acidic polysaccharide–detergent complex is thus precipitated on the surface of cellulose from which it is readily eluted by using increasing concentrations of salts such as $MgCl_2$ (*334,335*).

A vast literature exists on the biologically significant tricomplex systems of acidic polysaccharides–protein–microions. Earlier literature is summarized in Bugenberg de Jong's monograph (*324*). Since then three main paths have been followed in the investigations.

(1) Interactions of acidic polysaccharides with proteins which largely can be explained on an exclusion volume principle. In these cases the sum of the partial osmotic pressures was less than the total osmotic pressure, and the sedimentation phenomena of the tricomplex system (hyaluronate or chondroitin sulfate–protein–salt) similarly demonstrated a sieving effect (*336–342*). Such model systems, in the opinion of Laurent (*343*), can explain the transport phenomena observed in connective tissues. The connective tissues composed of acidic polysaccharide–protein networks allow the transport of macromolecules, such as globular proteins, either through the network compartment or intercompartment channels. The first kind of transport has its analogue in sedimentation, electrophoretic, and hydrodynamic studies on model acidic polysaccharide systems. The second kind of transport has its analogue in the gel filtration studies of macromolecules on acidic polysaccharide gel columns.

(2) At lower pH's specific electrostatic interactions occur which may show up as a separate electrophoretic component and which can be caused to flocculate (*147,344–346*). The shape of the aggregate gels or flocculates is dependent upon the electrostatic environment (*347*). The large number of precipitation reactions studied on chondromucoprotein–protein systems (*147,348,349*), heparin–fibrinogen (*350–353*) and carrageenan and gelatin (*354*) all indicate that the interaction is due to the interaction of the oppositely charge molecules. The stoichiometric relation of polyanion and protein is determined by the net positive charge on the protein and not the total number of positive charges. The stoichiometry is strongly affected by microcations and pH, both probably acting the same way; namely, by removing the charges from the anionic groups. Lately, however, evidence has appeared in the literature indicating that the interaction is more complex than originally thought. Many acidic polysaccharides gave precipitin reactions with aqueous solutions of gelatin. The

complex nature of the interaction forces was demonstrated by the fact that these precipitin reactions were inhibited by 0.145 N NaCl, 1.0 M urea, and 0.3 M guanidine HCl (*355*).

(3) There are also a number of soluble acidic polysaccharide–protein complexes which may show up in the reversible inhibition of the enzyme (*122,356–359*). The nature of such soluble complexes have been recently studied with chromatography, electrophoresis, and ultracentrifugation. The interaction here is relatively weak since the over-all charge on both macromolecules was negative but the cooperative effect of a number of sterically located positive charges caused the association. These soluble complexes are the intermediate type between the strictly steric type of interaction (sieve effect) the precipitation reactions (*360*).

REFERENCES

1. H. Muir, *Biochem. J.*, **69**, 195 (1958).

2. E. Buddecke, W. Kröz, and E. Lanka, *Z. Physiol. Chem.*, **331**, 196 (1963).

3. L. Rodén, J. D. Gregory, and T. C. Laurent, *Fed. Proc.*, **22**, 413 (1963).

4. B. Anderson, P. Hoffman, and K. H. Meyer, *J. Biol. Chem.*, **240**, 156 (1965).

5. U. Lindahl and L. Rodén, *Biochem. Biophys. Res. Commun.*, **17**, 254 (1964).

6. P. G. Johanssen, R. D. Marshall, and A. Neuberger, *Biochem. J.*, **236**, 2452 (1961).

7. V. Bogdanov, E. Kaverzneva, and A. Andreseva, *Biochim. Biophys. Acta*, **65**, 168 (1962).

8. K. Meyer, *Cold Spring Harbor Symp. Quant. Biol.*, **6**, 91 (1938).

9. W. Pigman and R. M. Goepp, Jr., *Chemistry of Carbohydrates*, Academic, New York, 1948.

10. H. Gibian, *Mucopolysaccharide and Mucopolysaccharidasen*, Deuticke, Vienna, 1959.

11. R. W. Jeanloz, in *Comprehensive Biochemistry*, Vol. 5, (M. Florkin and E. H. Stotz, eds.), Elsevier, Amsterdam, 1963.

12. J. S. Brimacombe and J. M. Webber, *Mucopolysaccharides*, Elsevier, Amsterdam, 1963.

13. K. Meyer and J. W. Palmer, *J. Biol. Chem.*, **107**, 629 (1934).

14. K. H. Meyer, E. A. Davidson, A. Linker, and P. Hoffman, *Biochim. Biophys. Acta*, **21**, 506 (1956).

15. E. A. Balazs, T. C. Laurent, U. B. G. Laurent, M. H. DeRoche, and D. M. Bunney, *Arch. Biochem. Biophys.*, **81**, 464 (1959).

16. K. Meyer, *Physiol. Rev.*, **27**, 335 (1947).

17. B. Weissmann and K. Meyer, *J. Amer. Chem. Soc.*, **74**, 4729 (1952); **76**, 1753 (1954).

18. R. W. Jeanloz and H. M. Flowers, *J. Amer. Chem. Soc.*, **84**, 3030 (1962); *Biochemistry*, **3**, 123 (1964).

19. K. Meyer, A. Linker, E. A. Davidson, and P. Weissmann, *J. Biol. Chem.*, **205**, 611 (1953).

20. G. Fischer and D. Boedeker, *Ann. Chem.*, **117**, 111 (1861).

21. P. A. Levene, *Hexosamines and Mucoproteins*, Longmans, Green, London, 1925.

22. M. L. Wolfrom and B. O. Juliano, *J. Amer. Chem. Soc.*, 82, 1673 (1960); 82, 2588 (1960).
23. E. A. Davidson and K. Meyer, *J. Amer. Chem. Soc.*, 76, 5686 (1954).
24. F. A. Bettelheim and D. E. Philpott, *Nature*, 188, 654 (1960).
25. F. A. Bettelheim, *Biochim. Biophys. Acta*, 83, 350 (1964).
26. K. Meyer and J. W. Palmer, *J. Biol. Chem.*, 114, 689 (1936).
27. S. F. D. Orr, *Biochem. Biophys, Acta*, 14, 173 (1954).
28. M. B. Mathews, *Nature*, 181, 421 (1958).
29. P. Hoffman, A. Linker, and K. H. Meyer, *Biochim. Biophys. Acta*, 30, 184 (1958).
30. O. Furth and T. Bruno, *Biochem. Z.*, 294, 153 (1937).
31. T. Soda, F. Egami, and P. Harigome, *J. Chem. Soc., Japan*, 51, 43 (1940).
32. S. Suzuki, *J. Biol. Chem.*, 235, 3580 (1960).
33. M. B. Mathews and M. Inouye, *Biochem. Biophys. Acta*, 53, 509 (1961).
34. K. Meyer and E. Chaffee, *J. Biol. Chem.*, 138, 491 (1941).
35. T. J. Stoffyn and R. W. Jeanloz, *J. Biol. Chem.*, 235, 2507 (1960).
36. R. W. Jeanloz, P. J. Stoffyn, and M. Tremege, *Fed. Proc.*, 16, 201 (1957).
37. P. Hoffman, A. L. Linker, V. Lippman, and K. H. Meyer, *J. Biol. Chem.*, 235, 3066 (1960).
38. L. Fransson and L. Rodén, *J. Biol. Chem.*, 242, 4161 (1967).
39. L. Fransson and L. Rodén, *J. Biol. Chem.*, 242, 4170 (1967).
40. L. Fransson, *Biochim. Biophys. Acta*, 156, 311 (1968).
41. J. McLean, *Amer. J. Physiol.*, 41, 250 (1916).
42. G. J. Durant, H. Hendrickson, and R. Mongomery, *Arch. Biochem. Biophys.*, 99, 418 (1962).
43. A. B. Foster, R. Harrison, T. D. Inch, M. Stacey, and J. M. Webber, *J. Chem. Soc.*, 1963, 2279.
44. J. E. Jorpes, H. Boström, and V. Mutt, *J. Biol. Chem.*, 183, 607 (1950).
45. K. H. Meyer and Z. E. Schwartz, *Helv. Chim. Acta*, 33, 1651 (1950).
46. G. Nomine, R. Bucourt, and P. Bertin, *Bull. Soc. Chim. Fr.*, 1961, 561.
47. I. Danishefsky, H. B. Eiber, and A. H. Williams, *J. Biol. Chem.*, 238, 2895 (1963).
48. M. L. Wolfrom, J. R. Vercellotti, H. Tomomatsu, and D. Horton, *J. Org. Chem.*, 29, 540 (1964).
49. Z. Yosizawa, *Biochem. Biophys. Res. Commun.*, 16, 336 (1964).
50. T. Kotoku, Z. Yosizawa, and F. Yamauchi, *Arch. Biochem. Biophys.*, 120, 553 (1967).
51. J. E. Jorpes and S. Gardell, *J. Biol. Chem.*, 176, 267 (1948).
52. A. Linker, P. Hoffman, P. Sampson, and K. H. Meyer, *Biochim. Biophys. Acta*, 29, 443 (1958).
53. A. Linker and B. P. Sampson, *Biochim. Biophys. Acta*, 43, 366 (1960).
54. J. A. Cifonelli and A. Dorfman, *Biochim. Biophys. Research Commun.*, 4, 328 (1961).
55. J. Knecht, J. A. Cifonelli, and A. Dorfman, *J. Biol. Chem.*, 242, 4652 (1967).
56. J. A. Cifonelli and L. Rodén, *Biochim. Biophys. Acta*, 165, 553 (1968).
57. K. H. Meyer, A. Linker, E. A. Davidson, and B. Weissman, *J. Biol. Chem.*, 205, 611 (1953).
58. S. Hirano, P. Hoffman, and K. Meyer, *J. Org. Chem.*, 26, 5064 (1961).
59. N. Seno, K. H. Meyer, B. Anderson, and P. Hoffman, *J. Biol. Chem.*, 240, 1005 (1965).
60. V. P. Bhavanandan and K. H. Meyer, *Science*, 151, 1404 (1966).

61. V. P. Bhavanandan and K. H. Meyer, *J. Biol. Chem.*, 243, 1052 (1968).
62. J. A. Cifonelli, A. Saunders, and J. I. Gross, *Carbohyd. Res.*, 3, 478 (1967).
63. J. W. Lash and M. W. Whitehouse, *Biochem. J.*, 74, 351 (1960).
64. V. E. Shashoua and H. Kwart, *J. Amer. Chem. Soc.*, 81, 2899 (1959).
65. K. Iida, *J. Biochem.* (Tokyo), 54, 181 (1963).
66. S. Hunt and F. R. Jevons, *Biochem. J.*, 97, 701 (1965).
67. S. Hunt and F. R. Jevons, *Biochim. Biophys. Acta*, 101, 214 (1965).
68. S. Hunt and F. R. Jevons, *Biochem. J.* 98, 522 (1966).
69. Z. I. Kertesz, *The Pectic Substances*, Wiley(Interscience), New York, 1951.
70. S. S. Bhattacharjee and T. E. Timall, *Can. J. Chem.*, 43, 758 (1965).
71. A. J. Barrett and D. H. Northcote, *Biochem. J.*, 94, 617 (1965).
72. W. L. Nelson and L. H. Cretcher, *J. Amer. Chem. Soc.*, 51, 1914 (1929).
73. E. L. Hirst, J. K. N. Jones, and W. O. Jones, *J. Chem. Soc.*, 1939, 1880.
74. E. Hirst and D. A. Rees, *J. Chem. Soc.*, 1965, 1182.
75. A. Haug, B. Larsen, and O. Smidsrød, *Acta Chem. Scand.*, 21, 691 (1967).
76. E. Hirst, W. Mackie, and E. Percival, *J. Chem. Soc.*, 1965, 2958.
77. B. Larsen, A. Haug, and T. J. Painter, *Acta Chem. Scand.*, 20, 219 (1966).
78. B. Larsen, *Acta Chem. Scand.*, 21, 1395 (1967).
79. A. N. O'Neill, *J. Amer. Chem. Soc.*, 77, 6324 (1955).
80. T. C. S. Dolan and D. A. Rees, *J. Chem. Soc.*, 1965, 3534.
81. A. J. Pernas, O. Smidsrød, B. Larsen, and A. Haug, *Acta Chem. Scand.*, 21, 98 (1967).
82. D. B. Smith and W. H. Cook, *Arch. Biochem. Biophys.*, 45, 232 (1953).
83. D. E. Timell, *Advan. Carbohyd. Chem.*, 19, 247 (1964); 20, 409 (1965).
84. G. O. Aspinall, E. L. Hirst, and R. S. Mahomed, *J. Chem. Soc.*, 1954, 1764.
85. G. S. Dutton and F. Smith, *J. Amer. Chem. Soc.*, 78, 2505 (1956).
86. D. Horton and M. L. Wolfrom, in *Comprehensive Biochemistry*, Vol. 5 (M. Florkin and E. H. Stotz, eds.), Elsevier, Amsterdam, 1963, p. 189.
87. D. M. W. Anderson, E. Hirst, and J. F. Stoddart, *J. Chem. Soc.*, C, 1967, 1476.
88. D. M. W. Anderson, E. Hirst, and J. F. Stoddart, *J. Chem. Soc.*, C, 1966, 1959.
89. D. M. W. Anderson and J. F. Stoddart, *Carbohyd. Res.*, 2, 104 (1966).
90. D. M. W. Anderson, I. C. M. Dea, and R. N. Smith, *Carbohyd. Res.*, 7, 320 (1968).
91. D. M. W. Anderson and I. C. M. Dea, *Carbohyd. Res.*, 5, 461 (1967).
92. D. M. W. Anderson and I. C. M. Dea, *Carbohyd. Res.*, 7, 109 (1968).
93. J. F. Stoddart and J. K. N. Jones, *Carbohyd. Res.*, 8, 29 (1968).
94. E. Janczura, H. R. Perkins, and H. J. Rogers, *Biochem. J.*, 80, 82 (1961).
95. G. T. Barry and W. F. Goebel, *Nature*, 179, 206 (1957).
96. G. T. Barry, *J. Exp. Med.*, 107, 507 (1958).
97. K. Heyns, G. Kiessling, W. Lindenberg, H. Paulsen, and M. W. Webster, *Chem. Ber.*, 92, 2435 (1959).
98. W. R. Clark, J. McLaughlin, and M. E. Webster, *J. Biol. Chem.*, 230, 81 (1958).
99. R. E. Reeves and W. F. Goebel, *J. Biol. Chem.*, 139, 511 (1941).
100. S. A. Barker, M. Stacey, and J. M. Williams, *Bull. Soc. Chim. Biol.*, 42, 1611 (1960).
101. J. K. N. Jones and M. B. Perry, *J. Amer. Chem. Soc.*, 79, 2787 (1957).
102. A. Linker and R. S. Jones, *Fed. Proc.*, 25, 410 (1966).
103. D. M. Carlson and L. W. Matthews, *Biochemistry*, 5, 2817 (1966).

104. A. Jeanes, C. A. Knutson, J. E. Pittsley, and P. R. Watson, *J. Appl. Polym. Sci.*, 9, 627 (1965).

105. I. R. Siddiqui, *Carbohyd. Res.*, 4, 277 (1967).

106. G. O. Aspinall, R. S. P. Jamieson, and J. F. Wilkinson, *J. Chem. Soc.*, 1956, 3483.

107. P. A. Sanford and H. E. Conrad, *Biochemistry*, 5, 1508 (1966).

108. H. E. Conrad, J. R. Bamburg, J. D. Epley, and T. J. Kindt, *Biochemistry*, 5, 2808 (1966).

109. E. Jansen, German Patent 332203 (1918).

110. J. K. Chowdhury, *Biochem. Z.*, 148, 76 (1924).

111. F. Höppler, *Chem. Ztg*, 66, 132 (1942).

112. J. V. Karabinos and M. Hindert, *Advan. Carbohyd. Chem.*, 9, 285 (1954).

113. J. F. Haskins, *Advan. Carbohyd. Chem.*, 2, 280 (1946).

114. C. R. Fordyce, *Advan. Carbohyd. Chem.*, 1, 309 (1945).

115. R. L. Whistler, *Advan. Carbohyd. Chem.*, 1, 279 (1945).

116. C. R. Ricketts, *Biochem. J.*, 51, 120 (1952).

117. E. L. Jackson and C. S. Hudson, *J. Amer. Chem. Soc.*, 59, 2049 (1937).

118. H. A. Rutherford, F. W. Minor, A. R. Martin, and M. J. Harris, *J. Res. Nat. Bur. Std.*, 29, 131 (1942).

119. E. C. Yakel and W. O. Kenyon, *J. Amer. Chem. Soc.*, 64, 121 (1942).

120. P. T. Mora, *J. Polym. Sci.*, 23, 345 (1957).

121. J. W. Wood and P. T. Mora, *J. Amer. Chem. Soc.*, 80, 3700 (1958).

122. P. T. Mora and B. G. Young, *Arch. Biochem. Biophys.*, 82, 6 (1959).

123. P. T. Mora, E. Merler, and P. Mauray, *J. Amer. Chem. Soc.*, 81, 5449 (1959).

124. J. J. O'Malley and R. H. Marchessault, *J. Polym. Sci., Part B*, 3, 685 (1965).

125. P. J. Flory, *Principles of Polymer Chemistry*, Cornell Univ. Press, Ithaca, New York, 1953.

126. R. A. Mock and C. A. Marshall, *J. Polym. Sci.*, 13, 263 (1954).

127. M. Nagasawa, M. Izumie, and I. Kagawa, *J. Polym. Sci.*, 37, 375 (1959).

128. Z. Alexandrowicz, *J. Polym. Sci.*, 56, 97, 115 (1962).

129. Z. Alexandrowicz and A. Katachalsky, *J. Polym. Sci., Part A*, 1, 3231 (1963).

130. F. Oosawa, *J. Polym. Sci., Part A*, 1, 1501 (1963).

131. K. A. Stacey, *Light Scattering in Physical Chemistry*, Academic, New York, 1956.

132. H. A. Schachaman, *Ultracentrifugation in Biochemistry*, Academic, New York, 1959.

133. T. C. Laurent, M. Ryan, and A. Peitruszkiewicz, *Biochem. Biophys. Acta*, 42, 476 (1960).

134. J. S. Johnson, K. A. Kraus, and G. Scatchard, *J. Phys. Chem.*, 58, 1034 (1954).

135. E. Braswell, *Biochim. Biophys. Acta*, 158, 103 (1968).

136. L. W. Nichol, A. G. Ogston, and B. N. Preston, *Biochem. J.*, 102, 407 (1967).

137. D. A. Yphantis, *Biochemistry*, 3, 297 (1964).

138. L. Sundelöf, *Ark. Kemi*, 29, 279 (1968).

139. G. V. Schultz, *Z. Phys. Chem.*, B43, 25 (1943).

140. H. J. Cantow and O. Fuchs, *Makromol. Chem.*, 83, 244 (1965).

141. C. E. Jensen and R. Durtoft, *Acta Chem. Scand.*, 8, 1659 (1954).

142. T. C. Laurent, *J. Biol. Chem.*, 216, 263 (1955).

143. B. S. Blumberg and A. G. Ogston, *Ciba Found. Symp. Chem. Biol. Mucopolysaccharides*, 1958, 22.

144. J. A. Christiansen and C. E. Jensen, *Acta Chem. Scand.,* 5, 849 (1951).

145. L. Varga, *J. Biol. Chem.,* 217, 651 (1955).

146. C. E. Jensen, *Acta Chem. Scand.,* 8, 292, 937 (1954).

147. M. B. Mathews, *Biochem. J.,* 96, 710 (1965).

148. M. B. Mathews, *Biochem. Biophys. Acta,* 35, 9 (1959).

149. M. B. Mathews, *Arch. Biochem. Biophys.,* 61, 367 (1956).

150. C. Tanford,E. Marler, E. Jury, and E. A. Davidson, *J. Biol. Chem.,* 239, 4034 (1964).

151. T. C. Laurent and A. Anseth, *Exp. Eye Res.,* 1, 99 (1961).

152. A. Anseth and T. C. Laurent, *Exp. Eye Res.,* 1, 25 (1961).

153. G. H. Barlow, L. J. Coen, and M. M. Mozen, *Biochem. Biophys. Acta,* 83, 272 (1964).

154. G. H. Barlow, N. D. Sanderson, and P. D. McNeill, *Arch. Biochem. Biophys.,* 94, 518 (1961).

155. F. Patat and H. G. Elias, *Naturwissenschaften,* 46, 322 (1959).

156. T. C. Laurent, *Arch. Biochem. Biophys.,* 92, 224 (1961).

157. J. Knecht and A. Dorfman, *Biochem. Biophys. Res. Commun.,* 21, 509 (1965).

158. The Svedberg and N. Gralen, *Nature,* 142, 261 (1938).

159. S. Säverborn, *Contributions to the Knowledge of Acid Polyuronides,* Almquist and Wiksells, Uppsala, 1945.

160. H. Owens, H. Lotzkar, T. H. Schultz, and W. D. Maclay, *J. Amer. Chem. Soc.,* 68, 1620 (1946).

161. O. Smidsrød and A. Haug, *Acta Chem. Scand.,* 22, 797 (1968).

162. W. H. Cook, R. C. Rose, and J. R. Colvin, *Biochem. Biophys. Acta,* 8, 595 (1952).

163. K. Arakawa, *Bull. Chem. Soc. Jap.,* 34, 1233 (1961).

164. A. Veis and D. N. Eggenberger, *J. Amer. Chem. Soc.,* 76, 1560 (1954).

165. H. B. Oakley, *Trans. Faraday Soc.,* 31, 136 (1935).

166. B. R. Record and M. Stacey, *J. Chem. Soc.,* 1948, 1561.

167. J. L. Strominger and J. M. Ghuysen, *Biochem. Biophys. Res. Commun.,* 12, 418 (1963).

168. W. Brown and D. Henley, *Makromol. Chem.,* 79, 68 (1964).

169. N. Schneider and P. Doty, *J. Phys. Chem.,* 58, 762 (1958).

170. W. Brown, D. Henley, and J. Öhman, *Ark. Kemi,* 22, 189 (1964).

171. R. L. Whistler and W. W. Spencer, *Arch. Biochem. Biophys.,* 95, 36 (1961).

172. Y. Tsang and T. E. Thompson, *J. Phys. Chem.,* 69, 4242 (1965).

173. W. M. Pasika and L. H. Cragg, *J. Polym. Sci.,* 57, 301 (1962).

174. W. Pigman and S. Rizvi, *Biochem. Biophys. Res. Commun.,* 1, 39 (1959).

175. W. Pigman, S. Rizvi, and H. Holley, *Biochem. Biophys. Acta,* 53, 254 (1961).

176. W. Pigman, S. Rizvi, and H. L. Holley, *Arthritis Rheumat.,* 4, 240 (1961).

177. G. Blix, *Acta Orthopaed. Scand.,* 18, 19 (1948).

178. L. Sundblad, *Acta Soc. Med. Upsaliensis,* 58, 113 (1953).

179. B. S. Blumberg, *Rheumatism,* 14, 37 (1958).

180. A. Guinier and G. Fournet, *J. Phys. Radium,* 8, 345 (1947).

181. O. Kratky, *Progr. Biophys.,* 13, 105 (1963).

182. S. Heines, O. Kratky, and J. Roppert, *Makromol. Chem.,* 56, 150 (1962).

183. S. S. Stivala, M. Herbst, O. Kratky, and I. Pilz, *Arch. Biochem. Biophys.,* 127, 795 (1968).

184. S. S. Stivala, L. Yuan, J. Ehrlich, and P. A. Liberti, *Arch. Biochem. Biophys.,* 122, 32 (1967).

185. W. G. McMillan and J. E. Mayer, *J. Chem. Phys.*, 13, 276 (1945).

186. C. Tanford, *Physical Chemistry of Macromolecules*, Wiley, New York, 1961.

187. B. N. Preston, M. Davies, and A. G. Ogston, *Biochem. J.*, 96, 449 (1965).

188. F. A. Bettelheim, Ph.D. Dissertation, University of California, 1956.

189. H. G. Reik and F. Gebert, *Z. Elektrochem.* 58, 458 (1954).

190. J. H. Hildebrand and R. L. Scott, *Solubility of Non-electrolytes*, 3rd ed., Reinhold, New York, 1950.

191. P. J. Flory and W. R. Krigbaum, *J. Chem. Phys.*, 18, 1086 (1950).

192. B. H. Zimm, W. H. Stockmayer, and M. Fixman, *J. Chem. Phys.*, 21, 1716 (1953).

193. T. A. Orofino and P. J. Flory, *J. Chem. Phys.*, 26, 1067 (1957).

194. A. Katchalsky and P. Spitnik, *J. Polym. Sci.*, 2, 432 (1947).

195. H. G. Spencer, *J. Polym. Sci.*, 56, 163 (1962).

196. H. Deuel, J. Solms, and H. Altermatt, *Vierteljahresschr. Naturforsch. Ges. Zuerich*, 98, 49 (1953).

197. R. Speiser, C. H. Hills, and C. R. Eddy, *J. Phys. Chem.*, 49, 328 (1945).

198. R. P. Newbold and M. A. Joslyn, *J. Ass. Offic. Agr. Chem.*, 35, 892 (1952).

199. I. Hirota, *Kogyo Kagaku Zasshi*, 64, 1262 (1961).

200. M. Rinaudo-Duhem, *C. R. Acad. Sci., Paris*, 258, 4042 (1964).

201. V. S. R. Roa and J. F. Foster, *Biopolymers*, 3, 185 (1965).

202. A. Haug, *Acta Chem. Scand.*, 15, 950 (1961).

203. J. T. G. Overbeek, *Bull. Soc. Chim. Belges*, 57, 252 (1948).

204. A. Katchalsky, N. Shavit, and H. Eisenberg, *J. Polym. Sci.*, 13, 69 (1954).

205. K. E. Kuettner and A. Lindenbaum, *Science*, 144, 1228 (1964).

206. K. E. Kuettner and A. Lindenbaum, *Biochim. Biophys. Acta*, 101, 223 (1965).

207. S. A. Rice and M. Nagasawa, *Polyelectrolyte Solutions*, Academic, New York, 1961.

208. I. M. Klotz, *J. Amer. Chem. Soc.*, 68, 1486, 2299 (1946).

209. M. B. Mathews, *Arch. Biochem. Biophys.*, 104, 394 (1964).

210. E. Buddecke and R. Drzenick, *Z. Physiol. Chem.*, 327, 49 (1962).

211. B. Carroll and H. C. Cheung, *J. Phys. Chem.*, 66, 2585 (1962).

212. M. D. Young, G. O. Phillips, and E. A. Balazs, *Biochim. Biophys. Acta*, 141, 374 (1967).

213. H. A. Benesi and J. H. Hildebrand, *J. Amer. Chem. Soc.*, 71, 2703 (1949).

214. E. A. Balazs, G. O. Phillips, and M. D. Young, *Biochim. Biophys. Acta*, 141, 382 (1967).

215. E. A. Balazs, J. V. Davis, G. O. Phillips, and D. S. Scheufele, *Biochem. Biophys. Res. Commun.*, 30, 386 (1968).

216. A. L. Stone and D. F. Bradley, *Biochim. Biophys. Acta*, 148, 172 (1967).

217. A. L. Stone, *Biochim. Biophys. Acta*, 148, 192 (1967).

218. J. E. Scott and I. H. Willett, *Nature*, 209, 985 (1966).

219. J. R. Dunstone, *Biochem. J.*, 85, 336 (1962).

220. A. Haug and O. Smidsrød, *Acta Chem. Scand.*, 19, 341 (1965).

221. E. Körös, H. Z. Remport, A. Lásztity, and E. Schulek, *Magy. Kem. F.*, 71, 369 (1965).

222. A. Haug and O. Smidsrød, *Nature*, 215, 757 (1967).

223. S. Salminen and K. Luomanmäki, *Biochim. Biophys. Acta*, 69, 533 (1963).

224. J. J. Farber and M. Schubert, *J. Clin. Invest.*, 36, 1715 (1957).

225. S. R. deGroot, *Thermodynamics of Irreversible Processes*, Wiley(Interscience), New York, 1951.

226. F. Akkerman, D. T. E. Pals, and J. J. Hermans, *Rec. Trav. Chim. Pays-Bas,* **71**, 56 (1952).

227. H. Fujita and T. Homma, *J. Polym. Sci.,* **15**, 277, (1955).

228. U. Lohmander and R. Strömberg, *Makromol. Chem.,* **72**, 143 (1964).

229. J. Schurz and H. Pippen, *Monatsh. Chem.,* **94**, 859 (1963).

230. P. J. Flory and T. G. Fox, Jr., *J. Amer. Chem. Soc.,* **73**, 1904 (1951).

231. R. C. Warner and M. Schubert, *J. Amer. Chem. Soc.,* **80**, 5166 (1958).

232. G. H. Barlow and L. J. Coen, *Biochem. Biophys. Acta,* **69**, 569 (1963).

233. A. R. Mathieson and M. R. Porter, *J. Polym. Sci.,* **21**, 483 (1956).

234. D. M. W. Anderson and K. A. Karamalla, *J. Chem. Soc., C,* **1966**, 762.

235. D. M. W. Anderson and S. Rahman, *Carbohyd. Res.,* **4**, 298 (1967).

236. K. O. Pederson, *J. Phys. Chem.,* **62**, 1282 (1958).

237. M. Smoluchowski, *Z. Phys. Chem.,* **92**, 129 (1918).

238. D. C. Henry, *Proc. Roy. Soc., Ser. A,* **133**, 106 (1931).

239. T. Foster and R. Pearce, *Can. J. Biochem.,* **39**, 1771 (1962).

240. L. G. Longsworth, in *Electrophoresis,* (M. Bier, ed.), Academic, New York, 1959.

241. H. Douglas and D. Shaw, *Trans. Faraday Soc.,* **53**, 512 (1957); **54**, 1748 (1958).

242. P. Doty and G. Ehrlich, *Ann. Rev. Phys. Chem.,* **3**, 91 (1952).

243. W. Ward, H. Swenson, and H. Owens, *J. Phys. Colloid Chem.,* **51**, 1137 (1947).

244. P. Hoffman, T. A. Mashburn, Jr., K. Meyer, and B. Anderson Bray, *J. Biol. Chem.,* **242**, 3799 (1967).

245. M. L. Wolfrom and F. A. H. Rice, *J. Amer. Chem. Soc.,* **69**, 2918 (1947).

246. Wm. Niedermeier and E. S. Gramling, *Carbohyd. Res.,* **8**, 317 (1968).

247. F. A. Bettelheim and A. Block, *Biochim. Biophys. Acta,* **165**, 405 (1968).

248. L. Varga and J. Gergely, *Biochim. Biophys. Acta,* **23**, 1 (1957).

249. G. Blix and O. Snellman, *Ark. Kemi, Mineral, Geol.,* **19A**, 32 (1945).

250. R. Brunish, J. W. Rowen, and S. R. Irvine, *Trans. Amer. Ophtal. Soc.,* **52**, 369 (1954).

251. D. Bourgoin, *J. Chim. Phys.,* **60**, 902, 911, 923 (1963).

252. K. J. Palmer and M. B. Hertzog, *J. Amer. Chem. Soc.,* **67**, 2122 (1945).

253. W. T. Astbury, *Nature,* **155**, 667 (1945).

254. F. A. Bettelheim, *J. Phys. Chem.,* **63**, 2069 (1959).

255. R. H. Marchessault and C. Y. Liang, *J. Polym. Sci.,* **59**, 357 (1962).

256. S. T. Bayley, *Biochim. Biophys. Acta,* **17**, 194 (1955).

257. H. S. Isbell and R. S. Tipson, *J. Res. Nat. Bur. Stand., A,* **64**, 171 (1960).

258. R. S. Tipson and H. S. Isbell, *J. Res. Nat. Bur. Stand., A,* **64**, 405 (1960); **65**, 31, 249 (1961).

259. H. G. Higgins, C. M. Stewart, and K. J. Harrington, *J. Polym. Sci.,* **51**, 59 (1961).

260. S. Krimm, *J. Polym. Sci., Part C,* **7**, 3 (1964).

261. R. S. Stein, *J. Polym. Sci.,* **28**, 83 (1958).

262. H. J. Marriman, *J. Polym. Sci.,* **39**, 461 (1959).

263. C. Y. Liang and R. H. Marchessault, *J. Polym. Sci.,* **37**, 385 (1959); **39**, 269 (1959); **43**, 85 (1960).

264. R. H. Marchessault and C. Y. Liang, *J. Polym. Sci.,* **43**, 71 (1960); **59**, 357 (1962).

265. F. G. Pearson, R. H. Marchessault, and C. Y. Liang, *J. Polym. Sci.,* **43**, 101 (1960).

266. K. H. Bassett, C. Y. Liang, and R. H. Marchessault, *J. Polym. Sci., Part A,* 1, 1687 (1963).

267. A. G. Lloyd, N. Tudball, and K. S. Dodgson, *Biochim. Biophys. Acta,* 52, 413 (1961).

268. S. F. R. Orr, *Biochim. Biophys. Acta,* 14, 173 (1954).

269. M. B. Mathews, *Nature,* 181, 421 (1958).

270. K. Onodera, S. Hirano, and N. Kashimura, *Carbohyd. Res.,* 1, 208 (1965).

271. F. R. Quinn and F. A. Bettelheim, *Biochim. Biophys. Acta,* 69, 544 (1963).

272. J. R. Nielson, *J. Polym. Sci., Part C,* 7, 19 (1964).

273. H. C. Hass, H. Husek, and L. D. Taylor, *J. Polym. Sci., Part A,* 1, 1215 (1963).

274. R. Cerf, J. Lang, and S. Caudau, *J. Polym. Sci., Part C,* 7, 163 (1964).

275. F. A. Bettelheim and R. S. Stein, *J. Polym. Sci.,* 27, 567 (1958).

276. O. Wiener, *Abhandl, Kgl. Sachs. Ges. Wiss. Math.-Phys. Kl.,* 32, 509 (1912).

277. K. Wiehrmann and W. Pilnik, *Experimentia,* 1, 330 (1945).

278. J. Palmer and M. B. Hertzog, *J. Amer. Chem. Soc.,* 67, 2122 (1945).

279. J. Palmer, R. C. Merrill, H. S. Owens, and M. Ballantyne, *J. Phys. Colloid Chem.,* 51, 710 (1947).

280. H. Thiele and G. Andersen, *Kolloid-Z.,* 140, 76 (1955).

281. J. L. Mongar and A. Wasserman, *J. Chem. Soc.,* 1952, 492.

282. C. Sterling, *Biochim. Biophys. Acta,* 26, 186 (1957).

283. B. Sylvén and E. J. Ambrose, *Biochim. Biophys. Acta,* 18, 857 (1955).

284. V. M. DerSarkissian and F. A. Bettelheim, *J. Polym. Sci., Part A,* 1, 725 (1963).

285. E. Russel, Thesis, Adelphi University, Garden City, New York, 1962.

286. A. Block and F. A. Bettelheim, *Biochim. Biophys. Acta,* 201, 69 (1970).

286a. F. A. Bettelheim, C. Sterling, and D. H. Volman, *J. Polym. Sci.,* 22, 303 (1956).

287. F. Kasahara, *Kogyo Kagaku Zasshi,* 62, 731 (1959); *Chem. Abstr.,* 57, 14039f (1962).

288. F. A. Betttelheim and S. H. Ehrlich, *J. Phys. Chem.,* 67, 1948 (1963).

289. F. A. Bettelheim and D. H. Volman, *J. Polym. Sci.,* 24, 445 (1957).

290. F. A. Bettelheim and D. H. Volman, *J. Polym. Sci.,* 24, 485 (1957).

291. A. B. D. Cassie, *Trans. Faraday Soc.,* 41, 458 (1954).

292. F. A. Bettelheim, *J. Colloid Interfac. Sci.,* 23, 301 (1967).

293. S. H. Ehrlich and F. A. Bettelheim, *J. Phys. Chem.,* 67, 1954 (1963).

294. I. Lubezky, F. A. Bettelheim, and M. Folman, *Trans. Faraday Soc.,* 63, 1794 (1967).

295. F. A. Bettelheim, Unpublished Data.

296. K. J. Palmer, T. M. Shaw, and M. Ballantyne, *J. Polym. Sci.,* 2, 318 (1947).

297. K. J. Palmer, R. C. Merrill, and M. Ballantyne, *J. Amer. Chem. Soc.,* 70, 570 (1948).

298. L. Kaufman and F. A. Bettelheim, *J. Polym. Sci.* in press.

299. T. C. Laurent, K. Hellsing, and I. Gelotte, *Acta Chem. Scand.,* 18, 274 (1964).

300. S. Miyamoto and T. Sasaki, *Bull. Chem. Soc. Japan,* 26, 228 (1953).

301. H. Thiele, *Naturwissenschaften,* 34, 123 (1947).

302. H. Thiele and G. Andersen, *Kolloid-Z.,* 140, 76 (1955); 142, 5 (1955); 143, 21 (1955).

303. H. Thiele and E. Schacht, *Z. Phys. Chem.,* 208, 42 (1957).

304. H. Thiele and E. Schacht, *Kolloid-Z.,* 161, 120 (1958).

305. H. Thiele and E. Schacht, *Kolloid-Z.,* 163, 2 (1959).

306. H. Thiele and K. Hallich, *Kolloid-Z.,* 151, 1 (1957).

307. H. Thiele and K. Hallich, *Z. Naturforsch., B,* 13, 580 (1958).

308. F. A. Bettelheim, *J. Polym. Sci., Part A-2,* 5, 1043 (1967).

309. H. Thiele, K. Plohnke, E. Brandt, and G. Moll, *Kolloid-Z.,* 182, 24 (1962).

310. H. Thiele, W. Joraschky, K. Plohnke, A. Wiechen, R. Wolf, and A. Wollmer, *Kolloid-Z.,* 197, 26 (1964).

311. F. A. Bettelheim and E. A. Balazs, *Biochim. Biophys. Acta,* 158, 309 (1968).

312. F. A. Bettelheim and M. Vinciguerra, *Biochim. Biophys. Acta,* 177, 259 (1969).

312a. M. Vinciguerra and F. A. Bettelheim, *Federation Proc.,* 29, 838 (1970).

313. O. Hector, T. Wittstruck, N. McNiven, and G. Lester, *Proc. Nat. Acad. Sci. U. S.,* 46, 783 (1960).

314. E. Balazs, A. Bothner-By, and J. Gergely, *J. Mol. Biol.,* 1, 147 (1959).

315. K. Arakawa, *Bull. Chem. Soc. Japan,* 35, 309 (1962); 57, 95g (1962).

316. J. S. Yudelson and R. E. Mack, *J. Polym. Sci., Part A,* 2, 4683 (1964).

317. E. A. Balazs, *Fed. Proc.,* 25, 1817 (1966).

318. H. Deuel and J. Solms, *Advan. Chem. Ser.,* 11, 62 (1954).

319. H. Owens, H. Swenson, and T. H. Schultz, *Advan. Chem. Ser.,* 11, 10 (1954).

320. E. Pippen, T. H. Schultz, and H. Owens, *J. Colloid Sci.,* 8, 97 (1953).

321. H. Harvey, *Chem. Ind.,* 7, 29 (1960).

322. A. A. Morozov, V. P. Rudi, and E. Z. Shifris, *Nauk. Zap. Chernivets'k. Univ.,* 33, 7 (1959); *Chem. Abstr.,* 57, 10294i (1962).

323. G. Baker, *Advan. Food Res.,* 1, 395 (1948).

324. H. G. Bungenberg de Jong, in *Colloid Science,* Vol. II (H. R. Kruyt, ed.), Elsevier, Amsterdam, 1949, Chap. X.

325. E. A. Balázs and J. A. Szirmai, *J. Histochem. Cytochem.,* 6, 278, 416 (1958).

326. B. Sylvén, *Quart. J. Microsc. Sci.,* 95, 327 (1954).

327. A. Hasson, and C. Chagas, *Comparative Bioelectrogenesis,* Elsevier, Amsterdam, 1960.

328. S. Ehrenpreis and M. G. Kellock, *Biochim. Biophys. Acta,* 45, 525 (1960).

329. S. Ehrenpreis, *Georgetown Med. Bull.,* 16, 148 (1963).

330. J. E. Scott, *Biochim. Biophys. Acta,* 18, 428 (1955).

331. J. E. Scott, *Biochem. J.,* 81, 418 (1961).

332. J. E. Scott, *Biochem. J.,* 84, 270 (1962).

333. J. E. Scott, *Biochem. J.,* 99, 3p (1966).

334. C. A. Antonopoulos, E. Borelius, S. Gardell, B. Mamnström, and J. E. Scott, *Biochim. Biophys. Acta,* 54, 213 (1961).

335. S. Schiller and A. Dorfman, *Nature,* 185, 111 (1960).

336. A. G. Ogston and C. F. Phelps, *Biochem. J.,* 78, 827 (1961).

337. A. G. Ogston, *Arch. Biochem. Biophys,. Suppl.* 1, 39 (1962).

338. T. C. Laurent and A. G. Ogston, *Biochem. J.,* 89, 249 (1963).

339. T. C. Laurent and H. Persson, *Biochim. Biophys. Acta,* 78, 360 (1963).

340. T. C. Laurent and A. Pietruszkiewicz, *Biochim. Biophys. Acta,* 49, 258 (1961).

341. T. C. Laurent, *Biochem. J.,* 93, 106 (1964).

342. B. R. Gerber and M. Schubert, *Biopolymers,* 2, 259 (1964).

343. T. C. Laurent, *Fed. Proc.,* 25, 1128 (1966).

344. M. G. Blair, W. Pigman, and H. L. Holley, *Arthritis Rheumat.* 4, 612 (1961).

345. D. Platt, W. Pigman, and H. L. Holley, *Arch. Biochem. Biophys.,* 79, 224 (1959).

346. E. Gramling, W. Niedermeier, H. L. Holley, and W. Pigman, *Biochim. Biophys. Acta,* 69, 552 (1963).

347. F. A. Bettelheim and D. E. Philpott, *Biochim. Biophys. Acta,* 34, 124 (1959).

348. A. J. Anderson, *Biochem. J.,* 88, 460 (1963); 94, 401 (1965).

349. A. J. Anderson, *Nature,* 207, 408 (1965).

350. R. T. Smith and R. W. Von Korff, *J. Clin. Invest.,* 36, 596 (1957).

351. H. C. Godal, *Scand. J. Clin. Lab. Invest.,* 13, 550 (1961).

352. K. W. Walton, *Brit. J. Pharmacol.,* 9, 1 (1955).

353. P. Bernfeld and J. S. Nisselbaum, *Fed. Proc.,* 15, 220 (1956).

354. E. A. MacMullan and F. R. Eirich, *J. Colloid Sci.,* 18, 526 (1963).

355. E. E. Woodside, G. F. Trott, R. J. Doyle, and C. W. Fishel, *Carbohyd. Res.,* 6, 449 (1968).

356. P. C. Spensley and H. J. Rogers, *Nature,* 173, 1190 (1954).

357. H. J. Rogers and P. C. Spensley, *Biochim. Biophys. Acta,* 13, 293 (1954).

358. E. Katchalski, A. Berger, and H. Newman, *Nature,* 173, 998 (1954).

359. M. Schubert and E. C. Franklin, *J. Amer. Chem. Soc.,* 83, 2920 (1961).

360. F. A. Bettelheim, T. C. Laurent, and H. Pertoft, *Carbohyd. Res.,* 2, 391 (1966).

CHAPTER 4

PHASE EQUILIBRIA IN SYSTEMS
OF INTERACTING POLYELECTROLYTES

Arthur Veis

DEPARTMENT OF BIOCHEMISTRY
NORTHWESTERN UNIVERSITY MEDICAL SCHOOL
CHICAGO, ILLINOIS

As a general rule, random chain polymer molecules are not miscible with small molecule solvents over the entire mixing composition range except in the situation where strong exothermic polymer–solvent interactions occurs. In a homogeneous polymer system the binary solvent–polymer temperature–composition phase diagram will consist of the binodal curve defining two liquid phases, one concentrated, the other dilute. However, at temperatures in the region of the consolute or critical value the "concentrated" phase may still retain much more solvent than solute on a volume or weight fraction basis as well as on a mole fraction basis

(*1*). The polymer concentration at the consolute temperature is generally quite low and will be a decreasing function of the molecular weight.

In ternary systems containing two dissimilar random chain polymer components and a single solvent component, and under conditions where each polymer is soluble in the solvent when in binary mixture, the usual state of affairs is that the system splits into two phases, each of which contains essentially only one of the polymer components (*2,3*). At very high molecular weights only a trivially small endothermic polymer–polymer intercation is sufficient to bring about the phase separation, and even polymers of very similar composition and structure are usually highly incompatible (*4*).

When the polymer–polymer interaction is exergonic, for whatever reason, the situation is dramatically altered. Phase separation still may occur and yield a concentrated and a dilute phase, but in this case both phases contain both polymers, with the concentrated phase usually containing the two polymers in some fixed ratio regardless of their initial mixing ratio. In strongly interacting systems, or where the configurational entropies of the polymers are small, the concentrated phase is a gel or precipitate. However, in a wide variety of cases where random chain polyions of opposite net charge are mixed, the concentrated phase is a very viscous liquid of moderate concentration. This phenomenon, discovered by Bungenberg de Jong and Kruyt (*5*) in 1930, was termed "complex coacervation" and is of special interest for several diverse reasons.

At the most fundamental level, the solution character of each phase allows one to explore the thermodynamics of the demixing process and the ensuing equilibrium states since statistical mechanical treatments of polymer solutions are available as guides. The corresponding treatment of a precipitate phase in general terms is not possible at this time.

In a quite different vein, Oparin (*6,7*) has suggested that the coacervation demixing process represents the primary ordering process by which mixtures of randomly formed prebiologic polymers were condensed into preprotoplasmic assemblies. The study of complex coacervation thus takes its place in the current investigations on the development of living systems via interactions between the random polymers produced in the primordial atmosphere of the earth. Indeed, it has been demonstrated (*8*) that enzymes can be selectively incorporated into coacervate droplets. Substrate added to the bulk solution diffused into the droplets and the converted products diffused out. The coacervate droplet, with inclusions, is thus an obvious model for the study of living systems.

More important than the question of the prebiologic organization of primordial "biopolymers" is the realization that living systems do consist of partitioned mixtures of polymeric polyelectrolytes, all in an essentially aqueous environment, and that functional units in many cases are distinct macromolecular complexes. Nucleoproteins and protein–polysaccharide complexes are two outstanding examples of such complexes. In most cases it is likely that biological selection of the most efficient processes has led to the development and utilization of rather specific pair-wise interactions, but an analysis of the general relationships in the demixing of complex mixtures of oppositely charged polyions is clearly fundamental to an understanding of many physiologically important systems. In this chapter we shall examine both the theory and phenomenology of the demixing processes resulting from nonspecific interactions of oppositely charged random chain polymeric polyions.

I. GENERAL CONSIDERATIONS ON THE THERMODYNAMICS OF PHASE SEPARATIONS IN RANDOM CHAIN POLYMER MIXTURES

A. Fundamental Equations

The free energy change, ΔG_M^a, for the athermal, ideal mixing of n_1 and n_2 moles of two pure liquid components, 1 and 2, respectively, at constant temperature, is given by

$$\frac{\Delta G_M^a}{RT} = n_1 \ln N_1 + n_2 \ln N_2 \tag{1}$$

for the case where components 1 and 2 have the same molar volume and N_1 and N_2 are mole fractions in the mixture. The corresponding ideal entropy of mixing, ΔS_M^a, is

$$\frac{\Delta S_M^a}{R} = -[n_1 \ln N_1 + n_2 \ln N_2] \tag{2}$$

When the two components are of unequal size, and in particular for the case of a random chain polymer mixed with solvent according to the Flory–Huggins (9,10) lattice model, Eqs. 1 and 2 become

$$\frac{\Delta G_M^{a'}}{RT} = \frac{\varphi_1}{r_1} \ln \varphi_1 + \frac{\varphi_2}{r_2} \ln \varphi_2 \tag{3}$$

$$\frac{\Delta S_M^{a'}}{R} = -\left[\frac{\varphi_1}{r_1} \ln \varphi_1 + \frac{\varphi_2}{r_2} \ln \varphi_2 \right] \tag{4}$$

where φ_i is the volume fraction of component i in the mixture and r_i the numbers of lattice sites occupied by i. The free energy and entropy of mixing functions are per mole of mixture. The reference states are pure solvent and pure random polymer. Figure 1 shows plots of Eq. 3 for the cases when $r_1 = r_2$ (Eq. 1) and $r_2 \gg r_1$. It is obvious that the effect

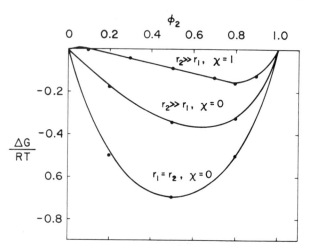

Fig. 1.   The free energy of mixing of polymer and solvent as a function of volume fraction, φ_2, of polymer and at various relative degrees of polymerization, r_2, and polymer–solvent interaction parameter χ.

of increasing solute size is to decrease the favorable free energy of mixing and markedly alter the symmetry of the free energy of mixing plot.   Thus, the entropy of transfer of a particular volume fraction or weight fraction of polymer from a dilute to a concentrated solution is much less unfavorable than the corresponding transfer of an identical weight fraction of a low molecular weight solute $(r_1 = r_2)$ from one solution to the other.

The mixing of unlike molecules is rarely athermal, so Eq. 3 must be modified by the inclusion of a heat of mixing term which, in general, will be proportional to the number of component 1–component 2 contacts. Interactions, whether favorable or unfavorable to 1–2 contacts, will cause deviations from random mixing and hence in the entropy of mixing term. This deviation will also be proportional to the number of 1–2 contacts. The net result of these similar concentration dependencies of heat and entropy of interaction is that one may include in Eq. 3 a composite correction term having the nature of a standard state free energy change due to interaction, $\Delta G_{M,i}$.   Thus

$$\Delta G_M = \Delta G_M^a + \Delta G_{M,i} = RT\left[\frac{\varphi_1}{r_1}\ln\varphi_1 + \frac{\varphi_2}{r_2}\ln\varphi_2 + \frac{\Delta G_{M,i}}{RT}\right] \quad (5)$$

Defining χ_{12} as the appropriate dimensionless proportionality constant per mole of component 1, it can be shown that

$$\Delta G_M = RT\left[\frac{\varphi_1}{r_1}\ln\varphi_1 + \frac{\varphi_2}{r_2}\ln\varphi_2 + \chi_{12}n_1\varphi_2\right] \quad (6)$$

where $\chi_{12} = BV_1/RT$, V_1 is the molar volume of sovent, component 1, and B is the van Laar interaction energy intensity characteristic of the 1–2 contacts. For heterogeneous mixtures all interactions between unlike pairs must be considered, Eq. 7.

$$\Delta G_M = RT \left[\frac{\varphi_1}{r_1} \ln \varphi_1 + \sum_{i=2} \frac{\varphi_i}{r_i} \ln \varphi_i + \sum_{\substack{i,j \\ \text{pairs}}} \chi_{ij} n_i \varphi_j \right] \tag{7}$$

B. Phase Separation

As long as a plot of ΔG_M vs φ_2 has, as in Fig. 1, a single minimum, the binary solution will be homogeneous. Positive values of the last right-hand term of Eqs. 6 or 7 will make ΔG_M less favorable for mixing and at some critical value, two minimal will appear, curve 3 of Fig. 1 for $r_2 \gg r_1$, $\chi_{12} = 1$, and solution mixtures at concentrations between these two minima will demix to form two phases whose compositions are given by the binodal tangent to the ΔG_M plot. At phase equilibrium the chemical potentials of each component in each phase must be equal, Eq. 8.

$$\left. \begin{array}{l} \mu_1^I = \mu_1^{II} \\ \mu_i^I = \mu_i^{II}, \ i = 2, 3, \cdots \end{array} \right\} \tag{8}$$

In all succeeding equations the Roman superscripts I and II will be used to denote dilute and concentrated phases, respectively. Where three or more coexisting phases are in equilibrium, superscript I will always refer to the dilute phase and II, III, . . . will refer to the concentrated phases in inverse order of their density or concentration. The critical conditions for phase separation, defining the consolute point, or the value of χ_{12} where an inflection point appears in the plot of ΔG_M vs φ_1, are given by

$$\left. \begin{array}{l} \left(\dfrac{\partial \mu_1}{\partial \varphi_2} \right)_{T,P} = 0 \\[3mm] \left(\dfrac{\partial^2 \mu_1}{\partial \varphi_2^2} \right)_{T,P} = 0 \end{array} \right\} \tag{9}$$

or the equivalent relationships

$$\left. \begin{array}{l} \left(\dfrac{\partial \mu_2}{\partial \varphi_2} \right)_{T,P} = 0 \\[3mm] \left(\dfrac{\partial^2 \mu_2}{\partial \varphi_2^2} \right)_{T,P} = 0 \end{array} \right\} \tag{10}$$

The critical value of χ_{12} for phase separation and the phase compositions can thus be determined if appropriate equations for μ_i can be obtained. Differentiation of Eq. 6, which describes the binary solution of solvent and homogeneous polymer, yields

$$\mu_1 - \mu_1^0 = RT\left[\ln\ (1\ -\ \varphi_2) + \left(1 - \frac{1}{r_2}\right)\varphi_2 + \chi_{12}\varphi_2^2\right] \tag{11}$$

and

$$\mu_2 - \mu_2^0 = RT\left[\ln\ \varphi_2 - (r_2 - 1) + r_2\varphi_2\left(1 - \frac{1}{r_2}\right) + \chi_{12}r_2(1\ -\ \varphi_2)^2\right] \tag{12}$$

These chemical potentials are not independent functions, but are related by the Gibbs–Duhem equation. Hence, since μ is directly accessible experimentally and since Eq. 11 is easier to deal with, it is usual to proceed by applying the conditions of Eq. 9 to Eq. 11 (*11*). The critical conditions arrived at in this way are

$$\varphi_{2,\mathrm{crit}} = \frac{1}{1 + r_2^{1/2}} \approx \frac{1}{r_2^{1/2}} \tag{13}$$

and

$$\chi_{\mathrm{crit}} = \frac{(1 + r_2^{1/2})^2}{2r_2} \approx \frac{1}{2} + \frac{1}{r_2^{1/2}} \tag{14}$$

It can be seen that for polymers of high molecular weight, $\varphi_{2,\mathrm{crit}}$ will always have a very low value, and χ_{crit} will have a value only slightly greater than 0.5.

C. Ternary Mixtures; Two Solutes and a Single Solvent Component

The ternary system of two dissimilar polymer components in a single solvent is of particular interest for the analysis of complex coacervation. We shall follow the treatment given by Scott (*12*) in some detail. Consider first the mixing of two polymers, components 2 and 3, in the absence of solvent, component 1. The chemical potentials corresponding to Eqs. 11 and 12 become

$$\mu_2 - \mu_2^0 = RT\left[\ln\ \varphi_2 + \left(1 - \frac{1}{r_2}\right)\varphi_2 + \chi_{23}\varphi_3^2\right] \tag{15}$$

and

$$\mu_3 - \mu_3^0 = RT[\ln\ \varphi_3 - (r_3 - 1)(1 - \varphi_3) + r_2\chi_{32}\varphi_2^2] \tag{16}$$

Assuming equal molecular weights for 2 and 3, $r_2 = r_3$ and $\chi_{23} = \chi_{32}$, and realizing that in this case the unit lattice site is the entire polymer so that $r_2 = r_3 = 1$, then

$$\mu_2 - \mu_2^0 = RT[\ln\ \varphi_2 + \chi_{23}\varphi_3^2] \tag{17}$$
$$\mu_3 - \mu_3^0 = RT[\ln\ \varphi_3 + \chi_{23}\varphi_2^2] \tag{18}$$

and the solution is identical with any regular binary mixture of liquids where all deviations from ideality are included in the van Laar term. Under those conditions $\chi_{23,crit} = 2$ and $\varphi_{2,crit} = \varphi_{3,crit} = \frac{1}{2}$. When $r_2 \neq r_1$ the critical equations become:

$$\chi_{23,crit} = \tfrac{1}{2}[1/(r_2)^{1/2} + 1/(r_3)^{1/2}]^2 \tag{19}$$

$$\varphi_{2,crit} = (r_2)^{1/2}/[(r_2)^{1/2} + (r_3)^{1/2}] \tag{20}$$

and

$$\varphi_{3,crit} = (r_3)^{1/2}/[(r_2)^{1/2} = (r_3)^{1/2}] \tag{21}$$

The parameters χ_{ij} as used thus far refer to the total contacts between interacting species and therefore are directly proportional to the molecular size of each. The interaction intensity per polymer segment in the binary polymer mixture described above then need only be a very small positive quantity to insure immiscibility of the unlike polymers. Alternatively, one may see that Eqs. 17–21 are the basis for the statement in the introduction that a favorable, exergonic interaction between polymers is necessary for their compatibility and such interactions tend to favor homogeneous solutions.

Now suppose that solvent ($r_1 = 1$, r_2, $r_3 \neq 1$) is added to the binary polymer mixture. The chemical potential expressions then take the form

$$\mu_1 - \mu_1^0 = RT\left[\ln \varphi_1 + (1 - \varphi_1) - \left(\frac{\varphi_2}{r_2} + \frac{\varphi_3}{r_3}\right) + (\chi_{12}\varphi_2 + \chi_{13}\varphi_3) \right.$$
$$\left. + (\varphi_2 + \varphi_3) - \chi_{23}\varphi_2\varphi_3/r_2 \right] \tag{22}$$

$$\mu_2 - \mu_2^0 = RT\left[\ln \varphi_2 + (1 - \varphi_2) - r_2\left(\varphi_1 + \frac{\varphi_3}{r_3}\right) \right.$$
$$\left. + (\chi_{21}\varphi_1 + \chi_{23}\varphi_3)(\varphi_1 + \varphi_3) - \chi_{13}r_2\varphi_1\varphi_3 \right] \tag{23}$$

$$\mu_3 - \mu_3^0 = RT\left[\ln \varphi_3 + (1 - \varphi_3) - r_3\left(\varphi_1 + \frac{\varphi_2}{r_2}\right) + (\chi_{31}\varphi_1 + \chi_{23}\varphi_2) \right.$$
$$\left. + (\varphi_1 + \varphi_2) - \chi_{12}r_3\varphi_1\varphi_2 \right] \tag{24}$$

It is worth pointing out that while χ_{12} and χ_{13} contain heat and entropy contributions, for positive values χ_{23} represents only the heat or energy of interaction. The use of these expressions is considerably simplified, but equally instructive, for the situation called the symmetrical mixing case by Scott in which $\chi_{12} = \chi_{13} = \chi$, $r_2 = r_3 = r$. With these restrictions, Eqs. 22–24 lead to result that, at phase equilibrium,

$$\varphi_1^I = \varphi_1^{II} \tag{25}$$

$$\varphi_2^I = \varphi_3^{II} \tag{26}$$

$$\varphi_3^I = \varphi_2^{II} \tag{27}$$

Obviously, the solvent and total polymer concentrations in each phase must be equal. The critical conditions are

$$\chi_{23,\text{crit}} = 2/(1 - \varphi_{1,\text{crit}})$$ (28)

and

$$\varphi_{2,\text{crit}} = \varphi_{3,\text{crit}} = \frac{(1 - \varphi_{1,\text{crit}})}{2}$$ (29)

If we set χ'_{23} equal to the interaction per polymer segment, or χ_{23}/r_2, then

$$\chi'_{23} = \frac{2}{r_2(1 - \varphi_{1,\text{crit}})}$$ (30)

and one can see that the interaction energies required for demixing are an order of magnitude below those required for demixing in the binary 1–2 solutions for polymers of comparable size. Comparing Eq. 29 and the prior observation that $\varphi_{2,\text{crit}} = \frac{1}{2}$ in the binary 2–3 case, the role of solvent is seen to be that of increasing the value of χ_{23} which can be tolerated without demixing. The polymer–solvent interactions, χ_{12} or χ_{13}, surprisingly have no effect on the phase separation due to endergonic polymer–polymer interactions as long as the concentrations are in the polymer–solvent miscibility range, i.e., χ_{12}, $\chi_{13} < 0.5$ in a 1–2 or 1–3 mixture.

D. Phase Separation in Ternary Mixtures where 2–3 Interactions are Exergonic

From the foregoing discussion it is clear that if $\Delta G_{M,i}$ is negative, χ_{23} of Eqs. 22–24 will also be negative and the critical value for demixing, Eq. 28 or 30, will never be achieved. Hence, phase separations where the attractively interacting polymers accumulate in separate phases will not occur. Phase separations do occur in such mixtures, however, with both polymers in both phases, and the substance of this review is to inquire into the modifications of the general treatment necessary to explain this type of demixing phenomenon.

Before going into detail, it is instructive to examine the general nature of the free energy of mixing plot when $\Delta G_{M,i}$ is negative because of 2–3 interactions. Assume a symmetric ternary mixture of components 2 and 3 such that $\chi_{12} = \chi_{13} < 0.5$, $r_2 = r_3 \gg r_1$, and $\varphi_2 = \varphi_3 = \frac{1}{2}(\varphi'_2)$. Equation 5 for the total free energy of mixing thus becomes

$$\Delta G_M = \Delta G_M^a + \Delta G_{M,i} = RT\left[(1 - \varphi'_2)\ln(1 - \varphi'_2) + \frac{\Delta G_{M,i}}{RT}\right]$$ (31)

For convenience, we set $\chi_{12} = \chi_{13} = 0$. Then $-\Delta G_{M,i}$ will be a function directly related to the total polymer concentration and which, in general, will be a monotonically increasing function of φ'_2. The first term (ΔG_M^a) in Eq. 31, the entropy of mixing, has the form shown in Fig. 2a. Since

ΔG_M^a and $\Delta G_{M,i}$ are additive, if $\Delta G_{M,i}$ is any linear function of φ_2' of whatever magnitude, ΔG_M vs φ_2' will have only a single minimum and the solution will remain homogeneous (Fig. 2a). If $\Delta G_{M,i}$ is an exponential function and is everywhere greater than the corresponding ΔG_M^a, phase separation will occur but will have the form of a precipitate of a solvated polymer phase of very high polymer concentration in equilibrium with an infinitely dilute solution. The coprecipitation behavior will persist for any exponential form of $\Delta G_{M,i}$ which becomes greater than ΔG_M^a at φ_2' smaller than its value at the maximum value of ΔG_M^a (Fig. 2b). When $\Delta G_{M,i}$ exceeds ΔG_M^a at φ_2' greater than the φ_2 at the maximum value of ΔG_M^a then, as indicated in Fig. 2c, two solution phases appear. However, in this case too, the concentrated phase will have a concentration on the order of $\varphi_2' \geqslant 0.5$ while the dilute phase may have $\varphi_2' \approx 0.1$–0.2.

In the coacervation systems to be examined the phase concentrations never approach these high values; on the contrary $(\varphi_2')^{I}$ is on the order of 0.01 and $(\varphi_2')^{II}$ is in the range of 0.03–0.05 over most of the demixing range. To achieve the requisite inflection in ΔG_M in this range, $\Delta G_{M,i}$ must have the concentration dependence illustrated in Fig. 2d. $\Delta G_{M,i}$ must increase rapidly at very low φ_2' and must be greater than the corresponding values of ΔG_M^a. At higher φ_2', $\Delta G_{M,i} < \Delta G_M^a$ and is relatively insensitive to φ_2'.

While one should not rely too heavily on such qualitative arguments, it is clear that when $\Delta G_{M,i}$ is negative only a very limited variety of concentration dependencies will produce liquid phase demixing rather than coprecipitation. Moreover, $\Delta G_{M,i}$ must have values of the same order of magnitude as ΔG_M^a.

There is an alternative approach to consider. The situations depicted in Fig. 2 all depend upon the adequacy of the random mixing equations (Eqs. 3 and 7) to describe the system with all deviations combined in $\Delta G_{M,i}$. It is possible to utilize some standard interaction energy expression, such as a Debye–Hückel term, and to attempt to modify ΔG_M^a, a procedure equivalent to incorporating a standard state change, or a change in definition of the nature of the "components" in the system. Both approaches have been taken by different investigators. Before discussing these treatments, however, we shall examine the experimental studies on complex coacervate systems in the next section.

II. THE PHENOMENOLOGY OF COMPLEX COACERVATION

A. The Basic System

In the usual study one takes a solution of a polycation (P$^+$), with its micro counterion (M$^-$) to preserve electrical neutrality, and mixes this

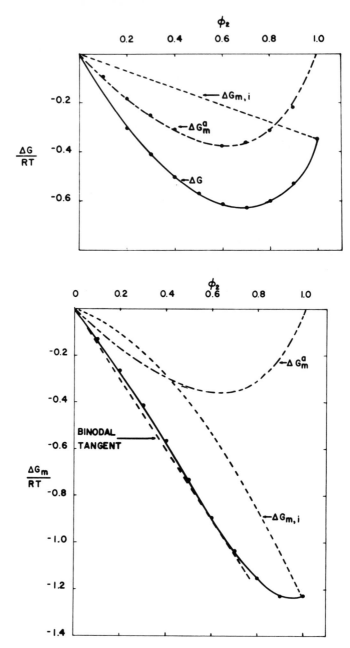

Fig. 2. The free energy of mixing as a function of the polymer volume fraction, φ_2, according to Eq. 31. (a) $\Delta G_{M,i}$, a linear function of φ_2, leading to homogeneous solutions. (b) $\Delta G_{M,i}$, an exponential function, everywhere greater than ΔG_M at

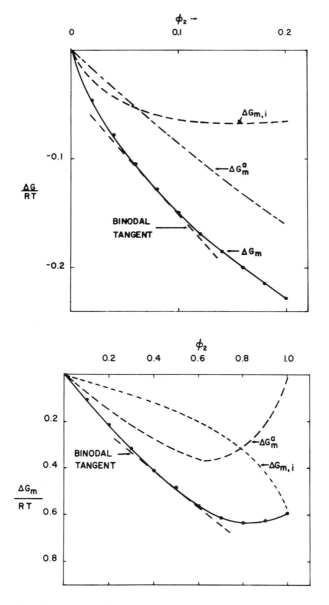

the same φ_2, leading to coprecipitation.   (c) $\Delta G_{M,i}$ and ΔG_M^a intersect at a φ_2 greater than the φ_2 for the maximum value of ΔG_M^a, leading to two liquid phases.   (d) $\Delta G_{M,i}$ greater than ΔG_M^a at some small value of φ_2, and attaining its maximum value $<\Delta G_M^a$ at some low value of φ_2, leading to two liquid phases with compositions similar to those found in complex coacervation situations.

with a solution of a polyanion (Q^-) and its micro counterion (N^+). The initial mixing concentrations and solution volumes may be varied at will but the case where the number of equivalents of polyion fixed charges are identical for P^+ and Q^- is of particular interest. The system is then designated as a "symmetrical" mixture. When conditions are such that phase separation occurs, both phases contain P^+ and Q^- and M^- and N^+, so that the system can in general be looked upon as a four component system of solvent, "polymer salt" P^+Q^-, "micro salt" N^+M^-, and P^+M^- or N^+Q^-, whichever is in excess. In the symmetrical mixture the system contains only three components. The dilute phase is called the "equilibrium liquid," and the concentrated phase the "coacervate." One measures the volume of each phase and the concentration of each component in each phase.

Three quantities of interest are: the "degree of coacervation," ρ,

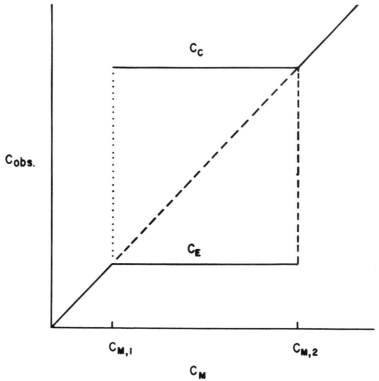

Fig. 3. Ideal two component phase separation. C_M, mixing concentration; C_{obs}, the phase concentration; C_c, the concentrated coacervate phase; C_E, the dilute equilibrium liquid phase; $C_{M,1}$, $C_{M,2}$, the lower and upper critical concentrations for phase separation, respectively.

defined as the fraction of polymer accumulated in the coacervate phase; the "enrichment," ϵ, defined as the ratio of polymer concentration in the concentrated and dilute phases; and the "intensity of coacervation," θ, defined by $\theta = \epsilon\rho$. Let us consider first the symmetric mixing case in which aqueous solutions of polyions are prepared in acid or base form so that the counter ions are H$^+$ and OH$^-$. Then on mixing equivalent amounts of P$^+$OH$^-$ and H$^+$Q$^-$, the final mixture will be neutral and can be represented as a two component system, P$^+$Q$^-$ and H$_2$O. The phase diagram for such a two component system, assuming ideal behavior, is depicted in Fig. 3 and the phase volumes in Fig. 4.   Setting

C_M = initial mixture concentration.
C_C = coacervate phase total concentration.
C_E = equilibrium liquid phase total concentration.
V_T, V_C, V_E = total volume, coacervate volume, and equilibrium liquid volume, respectively.

Then, in any case,

$$\rho = \frac{V_C C_C}{V_T C_M} = \frac{V_C C_C}{V_C C_C + V_E C_E} \tag{32}$$

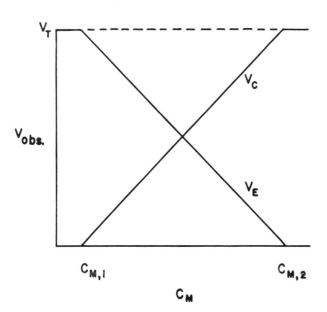

Fig. 4.   Volume relationships for the ideal two liquid component phase separation. V_{obs}, the phase volumes; V_T, the total mixture volume; V_E, the equilibrium liquid volume; V_c, the coacervate phase volume.

For the special case of ideal two component demixing, Eq. 32 reduces to

$$\rho = \frac{(1 - C_E/C_M)}{(1 - C_E/C_C)} = \frac{(1 - C_E/C_M)}{(1 - \epsilon)} \tag{33}$$

which has the relationship to C_M shown in Fig. 5.

The great majority of early studies were not carried out in the salt-free situation as indicated above. Each polyion was adjusted to the same pH by adding acid or base and then mixed at the proper equivalent concentration. In this situation the salt concentration, N^+M^-, increases as C_M increases. Hence, the phase diagrams were constructed at a variety of salt concentrations. As will be shown later in more detail, the addition of salt reduces the degree of coacervation. In this case, the phase diagram can be represented as in Fig. 6 (13). The equilibria depicted in Figs. 3, 4, and 5 all correspond to the line connecting $C_{M,1}$ and $C_{M,2}$ on the abscissa. The line marked "equivalence" denotes the phase relationships for the symmetrical mixtures where micro counterions are present. Upon mixing PM and NQ at $[PQ] = C'_M$, the mixture splits into two phases with concentrations C'_E and C'_C such that $C'_E > C_{M,1}$ and $C'_C < C_{M,2}$. The two phases contain, in general, slightly different concentrations of salt so that the tie lines are not parallel to the abscissa (14). When C'_M exceeds the value of C_M at which the equivalence line intersects the phase composition curve, C''_M, the salt concentration is so high that coacervation

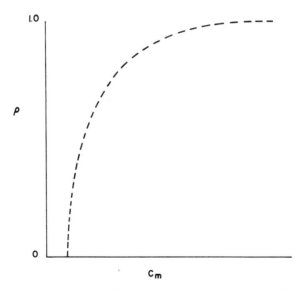

Fig. 5. The degree of coacervation, ρ, in the two component ideal case as a function of the mixing concentration.

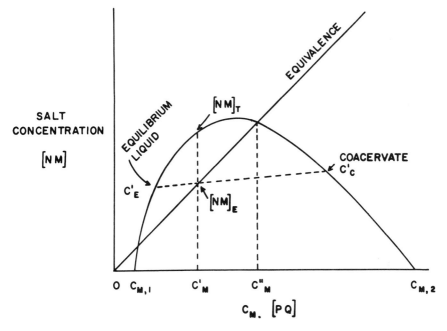

Fig. 6. The effect of salts on the phase diagram for complex coacervation. Salt is $[N^+M^-]$, polymer a mixture of P^+, Q^-. Other symbols as in the legend of Fig. 3.

is completely suppressed. This has been called "self-suppression of co-acervation" (*13*), but the term is not appropriate. We prefer to use the term "salt-suppression" for this case. If, at C'_M one adds additional micro-ion salt, NM, the phase compositions are gradually brought closer together until, at $[NM] = [NM]_T$, coacervation is again suppressed completely. $[NM]_T — [NM]_E$ is called the "salt tolerance" of the system. In the presence of microion salts, V_C and ρ, as well as C_E and C_C are complex functions of C_M and cannot be generalized as in Figs. 4 and 5 without detailed knowledge of the particular system.

The parameters used to describe complex coacervation can conveniently be divided between those which relate to the electrostatic interactions and those which relate principally to the backbone properties and molecular weights of the polymers. Indeed, the very extensive studies of Bungenberg de Jong and his many colleagues (see Ref. *15* for a comprehensive review) virtually ignored the polymeric nature of the polyions. It will be shown later that the electrostatic and chain conformation parameters are intimately related and this relationship should be kept in mind. How-ever, we shall discuss the experimental variables under these two broad classifications.

B. Electrostatic Effects

1. pH, Electrical Equivalence, and Charge Density

Bungenberg de Jong and Dekker (*15,16*) pointed out that a necessary condition for complex coacervation was a distinct charge opposition of the two polyions involved. They also observed there was an optimum mixing ratio at which ρ was maximized that corresponded to mixtures in which the polyions were present in exactly electrically equivalent amounts. The biopolymers used in coacervation studies generally contain weakly acidic or basic ionizable groups. In most cases the polyions are polyampholytes and the net charge densities are low. Thus, complex coacervation is very sensitive to both pH and relative mixing ratios.

For simplicity, most examples will be drawn from the system used in the author's laboratory. It is possible to prepare gelatins (random chain polypeptide polyampholytes) with different isoelectric points, depending on the pretreatment of the collagen stock. Under optimal conditions, gelatins with pI's of 5.0 and 9.0 may be obtained (*17*). These can be fractionated to yield preparations of similar weight-average molecular weight (*18*). Isoionic microion-free solutions (except for H^+ and OH^-) can be obtained by passing the gelatins over a mixed bed ion exchange column (*19*). Mixing solutions of such deionized gelatins provides a situation analogous to the symmetrical two component, salt-free case whose ideal behavior has been described above. Coacervates readily form in such mixtures at temperatures above the gelation temperature ($40°$) of the gelatins (*20,21*). The gelatin–gelatin system has the distinct advantage of involving polymers which are essentially identical in backbone properties and differing only in the presence or absence of a few amide groups. The charge–pH relationship for typical pI 5.0 and 9.0 gelatins is shown in Fig. 7, and the region of charge opposition is clearly defined. Upon mixing equal volumes of these two gelatins at equal weight concentrations the solution pH is 7.0 ± 0.2, although according to the titration data of Fig. 7 the correct equivalence point should be closer to pH 6.1. The titration data were obtained in the presence of salts, and this shifts the midrange titration region of the pI 9.0 gelatin somewhat (*20*). Coacervation does occur, in the absence of added salts, over the entire pH range of charge opposition (Fig. 8) although the coacervation is greatest, as measured by V_c, at the charge equivalence region. Salt suppression markedly narrows the pH range over which coacervation occurs. The equivalence point is clearly at pH 7.0–7.1.

An alternate way to study the equivalence problem is to maintain the pH, the total volume, and concentration of one polyion constant and to

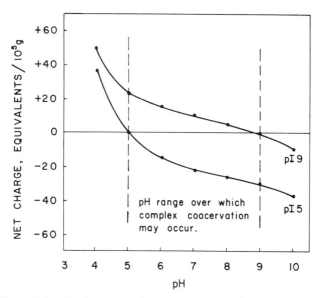

Fig. 7. The relationship between polymer component charge and the range over which coacervation can occur. Specific data are the titration curves of acid (pI 9)- and alkali (pI 5)-precursor gelatins in the mid-pH range. Data of Ames (*22*).

vary the concentration of the second polion. Data for a salt-free system containing denatured yeast deoxyribonucleic acid (DNA) and denatured pI 5.0 gelatin at 40° (*23*) are shown in Figs. 9 and 10. The gelatin concentration and total amount were held constant while the DNA concentration was varied. At the system pH, 3.2, exact electrical equivalence required a DNA/gelatin weight ratio of 0.71. The coacervate volume, V_c, went through a distinct maximum as the DNA/gelatin ratio was varied from 0.6 to 1.2. The coacervate concentration, on the other hand, increased over the entire mixing range and the fraction of total gelatin ($\rho_{gelatin}$) in the system gathered in the coacervate phase increased constantly up to a limiting value (Fig. 9) while ρ_{DNA} showed a distinct maximum. Most strikingly, these phase volume and concentration dependencies on mixing proportion all result from the tendency to maintain constant electrical equivalence in the concentrated phase, Fig. 10. Thus, in spite of wide variations in the polion ratios in the dilute phase, there is clearly polion–polion electrical equivalence in the coacervate. Excess polion and its counterion will not dilute or dissolve in the concentrated phase, as suggested by Voorn (*24*). Using mixtures of a pI 5 gelatin and a carboxylic polyanion, gum arabic, at pH 3.5, Bungenberg de Jong and Dekker (*16b*) found the ratio of gelatin to gum arabic in the coacervate phase to vary with the mixing concentration, although much less

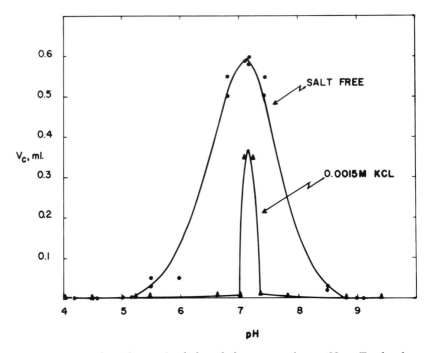

Fig. 8.   The pH dependence of gelatin–gelatin coacervation at 80°.   Total volume
was 10 ml, symmetric mixing with constant C_M.

markedly than the same ratio in the equilibrium liquid (Fig. 11).   These
data support the idea of a relative constancy of the equivalent composition of the concentrated phase.   The discrepancy probably arises from
the fact that Bungenberg de Jong used the salt forms of both polyions,
adjusting each solution to pH 3.5 with HCl.   In this case the gelatin and

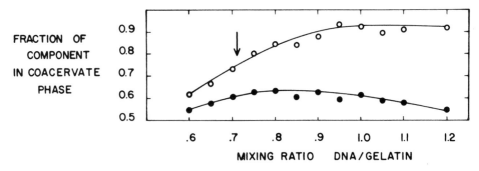

Fig. 9.   The fraction of each component accumulated in the coacervate phase, at
constant total mixture volume and total gelatin, as a function of the DNA/gelatin
weight ratio, 40°, pH 3.2.   The arrow indicates the electrostatic equivalence mixing
ratio   as   determined   from   titration   data.   Upper   curve: $\rho_{gelatin}$.   Lower   curve:
ρ_{DNA} (23).

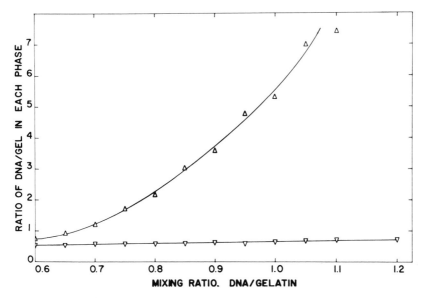

Fig. 10.   The ratio of DNA to gelatin in each phase as a function of the mixing ratio, conditions as in Fig. 9, showing the constancy of the coacervate phase polyion equivalence.   Upper curve: Equilibrium liquid.   Lower curve: The coacervate (23).

gum arabic solutions contained different amounts of micro counterion (gum arabic more than gelatin) and hence the microsalt concentration varied through the mixing range.   This point is emphasized here because all of their experiments are subject to the same uncertainty as to microion concentration.

The intensity of coacervation is directly related to the charge densities of the polyions.   As the charge density is increased the intensity of coacervation, θ, increases and the salt tolerance increases (25).   In some cases the high values of C_c and ρ in the coacervate phase bring about a decrease in V_C relative to a second system at lower charge density and θ (26).   Hence, the phase volume is not a reliable measure of coacervation intensity.   Michaels and Miekka (27) and Michaels et al. (28,29) have made use of the increased intensity and salt tolerance to prepare essentially insoluble polyanion–polycation complexes from polyions of very high charge density.   In this case (27,28) the equivalence of charged groups in the concentrated coprecipitated phase was very marked and excess polyions of either type, as well as excess counterions, were excluded from the complexed phase.

2. Salt-Suppression

As indicated schematically in Fig. 6, an increase in microion salt concentration suppresses coacervation, increasing C_E and more drastically

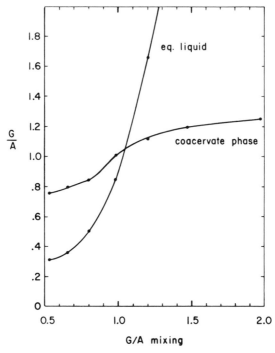

Fig. 11. The ratio of gelatin to gum arabic (G/A) in each phase in the G–A system at 40°, pH 3.5. The total polymer concentration in each case was 2% (*15*).

reducing C_C. A phase diagram constructed in a similar fashion for the deionized pI 5, pI 9 gelatin system is shown in Fig. 12 for the situation where C_M is held fixed and only the potassium chloride concentration is varied, corresponding to progression along the line $(C'_M — [NM]_T)$ of Fig. 6. Under the conditions of the measurement, the gelatins have a net charge of $\pm 15/10^5$ g or a net charge density, σ, of $\pm 0.015/$backbone residue; a very low σ, indeed.

The coacervate volume drops continuously over the $0–2.5 \times 10^{-3}$ M KCl range; however, C_C increases upon the addition of a minute amount of salt before dropping off sharply (high concentration branch, Fig. 12). The increase in C_M at low salt is not accompanied by an increase in ρ, however, so the composite intensity function θ, Fig. 13, decreases over the entire salt range. The quantities V_C, C_C, ρ, and θ are all complex functions of the salt concentration, but C_E appears to be a linear function of salt concentration.

If the polyion mixture is removed from the equivalence point, then the coacervate volumes may be increased upon the addition of small quantities of salt to the mixture before suppression takes place at higher salt con-

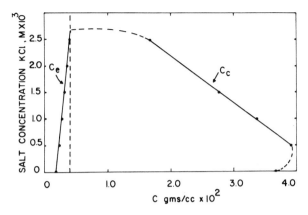

Fig. 12. The effect of salt on the composition of each phase at a constant mixing ratio. The gelatin–gelatin system, 80°, $C_M = 0.4\%$, $\sigma = \pm 0.015$.

centrations (30). As is to be expected, the valence of the microions is important, with the effectiveness of salts in the suppression of coacervation being in the order: 1–1 < 1–2, 2–1 < 1–3, 3–1. In an unsymmetrical mixture where polymeric cation is in excess, the initial enhancement in V_C is greatest when salts with polyvalent anions, e.g., 1–3, are used, a salt with a polyvalent cation 3–1 will only further suppress coacervation. The situation is reversed when polymeric anion is in excess, then 3–1 salts enhance and 1–3 salts suppress coacervation (30). These data indicate the role of preferential counterion binding in reducing the effective net charge on the polyions but show the ready replacement

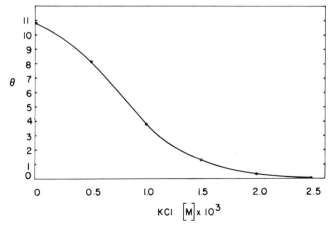

Fig. 13. The intensity of coacervation as a function of salt concentration. The conditions and system are the same as in the legend of Fig. 12.

of the bound microions with polyion when mixing is near the polyion equivalence point. Ion binding is probably the cause of the shift in C_c to higher values at very low salt concentrations noted in Fig. 12. The pI 9 gelatin has been shown to bind about 0.20 meq NaCl/g at pH 7 from 0.1 N NaCl solutions (20,31).

Michaels et al. (28) showed by conductivity studies on the system polystyrene sulfonate–(vinylbenzyl–trimethylammonium chloride) that virtually all the salt, even at $10^{-2} M$ equivalent concentrations of the counterions, was released from (or excluded from) the coprecipitate. Similar conductance studies on gelatin–gelatin coacervates (A. Veis and S. Mussell, unpublished results) showed, however, that even at very low charge densities the liquid coacervate phase had a conductance lower than that expected if the salts were present in an unbound state and distributed nearly uniformly in both phases. Since, by direct measurement of the chloride content of ashed coacervates in the gelatin–gum arabic system, Voorn (32) found chloride ion in both phases and in a ratio such that [Cl] coacervate/[Cl]$_{\text{eq. liq.}}$ > 1, it is likely that the low conductances of the coacervates are due to strong microion binding interactions. According to Voorn's data (32), the tie lines of Fig. 6 cannot be parallel to the abscissa and the plot in Fig. 12 of the phase concentration vs total salt concentration is slightly skewed.

C. Molecular Parameters

In view of the complicating effects of microion salts, and the fact that the various systems studied earlier in other laboratories were not characterized in terms of molecular weights or weight distributions, the ensuing

TABLE I

Approximate Molecular Parameters for Unfractionated
pI 5 and pI 9 Gelatins (21,33)

Gelatin				
Designation	Type	$M_W \times 10^{-5}$, in 2.0 M KCNS[a]	S^0_{40}, in 0.1 M KCl[a]	$[\eta]_{40}$, 0.2 M NaCl[b]
A	pI 9.0	3.3 ± 0.3	5.0	0.47
B	pI 5.0	3.3 ± 0.3	5.0	0.65

[a] Ref. 21. These gelatins gave broad single peaks with evidence of asymmetry on the leading edge. The values were computed from the maximum ordinate. Molecular weight by light scattering.

[b] Ref. 33.

TABLE II

Characterization Parameters for Fractionated Gelatins (*32*)

Gelatin		Ethanol recovery range alcohol H₂O ratio	Fraction designation	% initial gelatin	Fraction components			
Designation	Type				Designation	%	$S^0_{40,w}$	$[\eta]_{40}$
A	pI 9.0	2.3	A-I	24	A-I-1	40	11.6	1.42
					A-I-2	60	9.0	0.92
		2.3–2.7	A-II	27	A-II-1	86	7.8	0.72
					A-II-2	14	5.3	0.38
B	pI 5.0	1.90	B-I	22	B-I-1	96	8.2	0.85
					B-I-2	3	5.6	0.55
			B-II	28	B-II-1[a]	10	7.7	0.85
					B-II-2[a]	90	6.1	0.55

[a] These components are essentially identical with B-I-1 and B-I-2.

discussions will refer primarily to work on the symmetrical deionized pI 5–pI 9 gelatin systems used in the author's laboratory. These data thus refer to symmetrical mixtures of polyampholytes in the very low charge density range at ambient pH's always near neutrality. These are also the conditions of relevance to most of the biological problems to which we ultimately wish these data to apply.

The two gelatins were chosen because they had similar weight-average molecular weights and, hopefully, similar weight distributions. The average parameters for the gelatins are given in Table I. These values apply with reasonable precision to all of the gelatin preparations studied. Alcohol fractionation and retention of the high molecular weight fractions provided pauci-disperse preparations containing two-to-three components. The characterization parameters for the various fractions are given in Table II.

1. Concentration Dependence of Coacervation at Temperatures above the Gel Melting Point

a. *Unfractionated Gelatins. Self-Suppression of Coacervation*

The outstanding feature of coacervation in the salt-free mixtures of unfractionated gelatins is self-suppression, seen in both the phase composition diagram, V_C, and ρ plots of Figs. 14, 15, and 16. These data (*21,34*) are to be compared with the idealized plots of Figs. 3, 4, and 5. It is immediately evident that C_E never approaches a constant value, even over small mixing concentration ranges, whereas C_C strives to maintain a

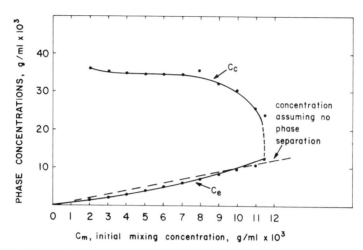

Fig. 14. Phase diagram for the gelatin–gelatin coacervation system, unfractionated gelatins, 40°. Gelatin weight-average molecular weights, 3.3×10^5.

constant value at low C_M. V_C reaches a maximum value at $C_M \approx 0.6\%$ and then rapidly falls to zero. At the same time ρ, which begins with a sharp rise at very low C_M, also drops off rapidly. While reminiscent of salt-suppression, these data were all obtained in the complete absence of microions other than H^+ or OH^-; hence, this phenomena has been termed "self-suppression."

In systems which show self-suppression the coacervation is very temperature dependent. The coacervation intensity decreases as the temperature is increased and the upper critical concentration for coacervation is lowered. In the unfractionated gelatin system, no coacervation could be detected at temperatures above 55°.

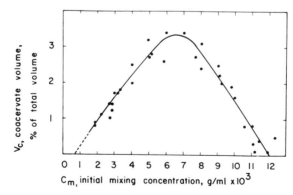

Fig. 15. Coacervate volumes for the system of Fig. 14.

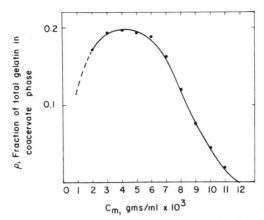

Fig. 16. Fraction of total gelatin in the coacervate phase for the system of Fig. 14.

b. *Molecular Weight Fractionation*

Light-scattering studies made on the separated phases showed clearcut evidence of a very marked molecular weight fractionation (*34*).   For example, the coacervate and equilibrium liquid phases, after disruption of any interaction effects by $2.0\,M$ KCNS, had weight-average molecular weights of 15×10^5 and 1.7×10^5, respectively, for a C_M of 0.6%, corresponding to the maxima in V_C and ρ of Figs. 15 and 16. M_W was 3.3×10^5 for the initial gelatin mixtures before coacervation (Table I). Thus, the coacervation process clearly places the higher weight components in the concentrated phase.   Intrinsic viscosities determined for both phases over the entire demixing range (*23,33*) emphasized that a very strong selectivity was maintained in the composition of the coacervate phase (Fig. 17).   At the low mixing concentration end, the coacervate phase intrinsic viscosity was constant and the greatest fraction was seen.   Where self suppression begins $(C_M \sim 0.5\%)$, the coacervate phase is diluted by lower weight components.   After the C_M corresponding to the maximum in V_C is surpassed, $[\eta]_C$ decreases rapidly.

Viscosity and light-scattering measurements in the absence of supporting electrolyte (*33,34*) were very difficult to carry out reproducibly. Nevertheless, two trends were evident.   The molecular weight in the equilibrium liquid was much above that of the gelatins in the equilibrium liquids after addition of salt (Table III) and M_W increased with increasing C_M.   On the other hand, the reduced viscosities of the equilibrium liquids were very low, considering the high molecular weights obtained.   These data were interpreted as indicating the presence of rather dense aggregates in the dilute phase.   The aggregates were readily dis-

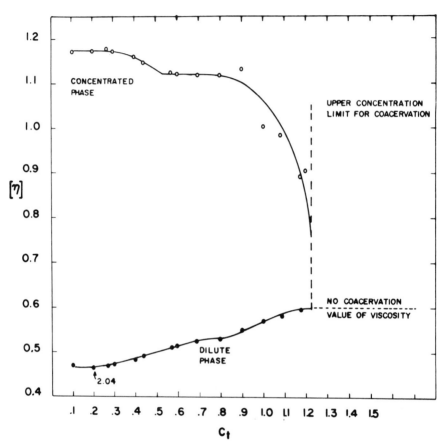

Fig. 17. Intrinsic viscosities of the gelatin and coacervate phases for unfractionated gelatins, same system as in Fig. 14. Viscosity measurements at 40° in 0.2 M NaCl (33).

TABLE III
Molecular Weights and Viscosities of Equilibrium Liquids from Unfractionated Gelatins at 40°

C_M, g/ml $\times$ 10³	$M_W \times 10^{-5}$ [a]	η_{red}, dl/g	η_{red}, dl/g, no coacervation
0.2	9.5	0.45	0.48
0.6	12.2	0.58	0.62
1.3[b]	20.2	0.82	0.82

[a] Weight-average values from light scattering (34).
[b] No coacervation at this mixing concentration (33).

sociated by the presence of very small amounts of microion salts.   However, the aggregates were stable to dilution in the absence of added microion salts.

It was not possible to do any light-scattering or viscosity measurements directly on the salt-free coacervate phase.   However, at 40° it was noted that generally the equilibrium liquid phase was turbid, whereas the coacervate phase was quite clear.

c. *Coacervation with Pauci-Disperse Gelatin Fractions*

The constant values of $[\eta]$ observed at low C_M in studies such as that illustrated in Fig. 17 led to the suggestion that particular components were being selected for incorporation into the coacervate.   Hence, self-suppression was thought to be due to the dilution of these components with lower molecular weight polyions.   Since ρ was on the order of 0.3 at its maximum value, fractionations were arranged as in Table II to collect the highest molecular weight one-third of both pI 5 and pI 9

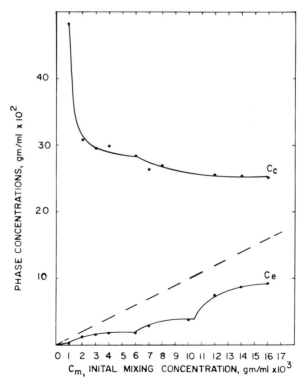

Fig. 18.   Phase diagram for the high molecular weight gelatin fraction coacervation system, A-I and B-I at 40°.   Compare with Fig. 14.

gelatins. Coacervation with these highest molecular weight fractions presented a markedly different behavior than in the previous, unfractionated case.

(*i*) *Concentration Dependence at Temperatures above the Gel Melting Point.* Figures 18, 19, and 20, to be compared with their counterparts Figs. 14, 15, and 16, show the variation of phase concentrations, V_C, and ρ as a function of the mixing concentration (*33*). This pauci-disperse system comes much closer to the predicted ideal behavior than does the less well fractionated mixture. Both C_C and C_E show C_M regions where an attempt is made to maintain constant phase composition, and it seems probable that a particular set of components is forming the coacervate phase in the 0.1–0.6% C_M region. In the C_M range from 0.6–1.1% a new set of components appears in the coacervate, and new plateau values of C_E and C_C are attained. Coacervation in this range is less intense than at lower concentration. Self-suppression can be seen in Fig. 20 in the sudden drop in ρ at $C_M \approx 0.6\%$ and again in the comparable region at $C_M > 1.0\%$ where further dilution of C_C takes place. These data lead directly to the conclusion that self-suppression of coacervation is related to molecular weight heterogeniety. Further, increasing the heterogeniety

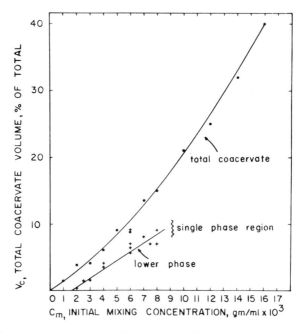

Fig. 19. Coacervate volumes for the system of Fig. 18. Compare with Fig. 15. Note that the coacervate is composed of two phases.

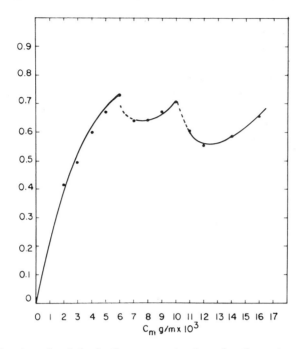

Fig. 20. Fraction of gelatin in the coacervate phase for the system of Fig. 18. Compare with Fig. 16.

of the concentrated phase components requires a reduction of the concentration of that phase. Hence, the concentration of each phase is clearly molecular-weight dependent. It should be noted that the initial, low C_M values of ρ are very nearly of the form predicted for the ideal symmetrical case (Fig. 5).

The upper critical mixing concentration at which coacervation occurs in the A-I-B-I system is on the order of $C_M = 2.2\%$. Measurements in this range become unreliable because it is very difficult to make a clean separation of the phases as C_C and C_E approach each other in value.

(*ii*) *The Homogeneity of the Coacervate Phase.* The coacervate concentration data described in the preceding section are correct in that they represent the average concentration of the total coacervate phase. However, as a general rule, in these pauci-disperse system mixtures the coacervate phase is itself composed of two phases. This can be discerned on close inspection by slight differences in the turbidities of the two phases and the appearance of a sharp refractive index gradient at the phase boundary. The more dense, concentrated phase is usually less turbid than the slightly less dense phase (*33,35*). The turbid phase might be considered as an artifact due to improper settling of the coacervate droplets. However, the appearance of the second phase, its volume, concentration, and temperature dependence is such that it appears to be a real rather than artifactual entity.

At 40° the two phases are restricted to a C_M range from 0.2–0.85% (Fig. 19). Single phase coacervates only are found at higher C_M values where, as described earlier, the total coacervate phase becomes more dilute. The mixing concentration range over which two coacervate phases appear is increased at elevated temperatures.

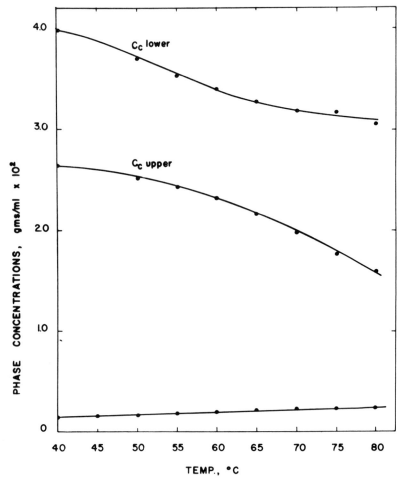

Fig. 21. The concentrations of the phases as a function of the temperature for the high molecular weight A-I and B-I fractions. $C_M = 0.3\%$. The lower curve is the concentration of the equilibrium liquid; the upper two curves, the two coacervate phase concentrations.

(*iii*) *Temperature Dependence of Coacervation.* In contrast to the unfractionated mixtures, the phase separation in A-I + B-I mixtures is not drastically suppressed by elevation of the temperature. It is therefore possible to study the temperature dependence of coacervation over a wide temperature range above the gel melting point. Since there are several different ranges of concentration dependence in different C_M ranges, the temperature dependence was measured (*33,35*) at several different values of C_M. The concentration ranges over which C_E was constant are of particular interest.

In every case two coacervate phases were noted. The volumes of the individual phases and their concentrations were determined as functions of T. Data obtained in the lowest constant C_E range, at $C_M = 0.3\%$, are shown in Figs. 21 and 22 (*35*). There is an obvious change in system behavior at temperatures above 60°. This is most marked upon examining the total coacervate volumes, Fig. 22, and is apparent at all mixing concentrations examined. As a preliminary to constructing thermodynamic equations and models for these systems, distribution coefficients, $\epsilon = C_C/C_E$, were determined for each phase at $C_M = 0.3$ and 0.8%. ϵ_1 is the distribution coefficient for the most concentrated phase; ϵ_2 is that for the less concentrated coacervate phase. Vant Hoff plots

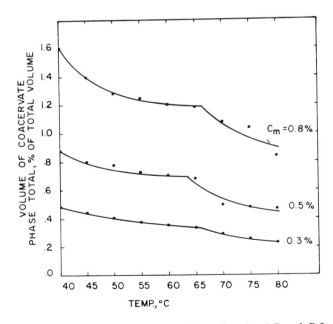

Fig. 22. Total volumes of the coacervate phases for the A-I and B-I system at various C_M and as a function of temperature.

of ln ϵ vs $1/T$ showed two linear branches in every case, with the change from low temperature branch to high occurring at $\sim 60°$. Apparent enthalpies for the transfer of polymer from dilute to a concentrated phase in the low T region for ϵ_1 and ϵ_2 at $C_M = 0.3$ and 0.8% were $-(4800 \pm 500)$ cal/mole. In each case, except for $\epsilon_2^{0.3\%}$, the enthalpies for the high temperature branch of both ϵ_1 and ϵ_2 were on the order of $-(2500 \pm 500)$ cal/mole. The plot of ln $\epsilon_2^{0.3\%}$ vs $1/T$ yielded a ΔH of -6500 cal/mole.

Figure 23 emphasizes a final point with regard to the temperature dependence of coacervation and its relationship to the mixing concentration. The most concentrated coacervate phase varies relatively little in dependence on C_M over the range $T \geqslant 55°$ and only slightly at lower T, but where it does differ this phase has a higher concentration the lower the value of C_M. In direct contrast, the more dilute coacervate phase is much more dependent on C_M and is most concentrated the higher the value of C_M. The very different temperature dependence of this dilute coacervate phase at $T > 55°$ at different values of C_M is also evident.

(*iv*) *Molecular Weight Fractionation.* Detailed studies on the fractionation from the pauci-disperse coacervation have not been completed because of the complications of multi-phase coacervates and the difficulties in cleanly separating the high molecular weight components of A-I and B-I after they are mixed. However, Veis et al. (*33*) did find that when A-I and B-I were mixed at the lowest feasible mixing concentration the coacervate phase was enriched in component A-I-1 (see Table II) with

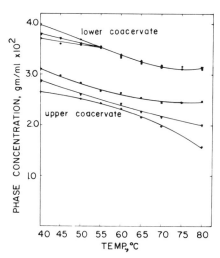

Fig. 23. The concentrations of the two coacervate phases as a function of temperature, A-I + B-I system. $C_M = 0.8\%$ (+), 0.5% (▲), and 0.3% (●).

respect to the initial A-I-1/A-I-2 ratio.   At the other extreme of very high C_M the A-I-1/A-I-2 ratio was nearly that of the original mixture. Veis and Weiss (unpublished results) have used carboxymethyl cellulose chromatography and electrofocusing pH gradient electrophoresis to separate the components of each phase after demixing.   Qualitatively it is evident that the higher weight components enter the most concentrated phase preferentially and that the lowest weight components are excluded from both coacervate phases where more than one coacervate exists. However, over the middle region coacervation range ($C_M = 0.8$–1.2%) the coacervate phases are not monodisperse.   These studies are still in progress.

2. Coacervation at Temperatures below the Gel Melting Point

The only studies on coacervation carried out at temperatures below the gel melting point of gelatin, including the very early experiments of Tiebackx (36), have utilized unfractionated gelatins (21,34,37) and hence are subject to all of the uncertainties caused by self-suppression and fractionation.   However, it is clear that below the melting temperature the coacervation directly parallels the course of collagenfold formation.   Figure 24 shows a plot of ρ vs T for an unfractionated $(A + B)$ gelatin mixture (21) at different C_M, representing the situation where coacervation just begins; where V_C and ρ have their maxima; and, at higher C_M, where coacervation is almost entirely suppressed at 40°.   The equilibrations were carried out in such a way that in each case the pI 5 and pI 9 gelatins were held at 40° before mixing.   The temperature was dropped to the equilibrium temperature immediately upon mixing.   After 30 min. the phases were coalesced and separated.   If one were to plot the specific optical rotation of a gelatin solution or any other measure of collagenfold formation on the same basis as the ρ in Fig. 24, the plots would have very similar shapes, with the "cooperative transition" region having the same width of 7–10° and the midpoint at about 25° (see Fig. 10, Ref. 38).

The question immediately arises as to whether the increase in coacervation is the result of the conformational change on fold-formation or of the increase in average molecular weight due to hydrogen-bonding interactions.   The latter seems most likely from the concentration dependence of the coacervation at low temperature (Fig. 24).   At low concentrations, where fold-formation is primarily intramolecular (39), ρ does not increase nearly as markedly with a decrease in temperature as at higher concentration where hydrogen bonding leads to intermolecular aggregates and substantial molecular weight increases.

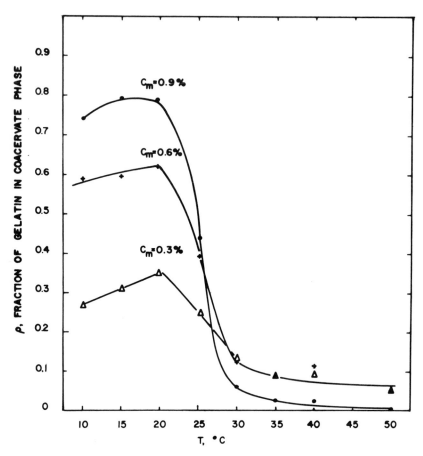

Fig. 24. The temperature dependence of coacervation in unfractionated mixtures of gelatins A and B at temperatures below the gelation temperature.

After a mixture has been coacervated at low temperature, heating back above the melting temperature without agitation did not cause an immediate dispersion of the coacervate phase, and it was observed that several coacervate phases were present (21). The number of phases depended upon both C_M and the equilibration temperature. Thus, even in the aggregation process some specific molecular weight or type selection was taking place. Light-scattering studies had indicated that the higher molecular weight components of A and B were sequestered in the coacervate phase at $C_M = 0.6\%$ and $40°$ (34). The gelatin in the equilibrium liquid phase from such a coacervate, with a weight-average molecular weight of 1.7×10^5, was collected, the temperature was dropped to $30°$, and the new phase equilibrium established. Molecular weights

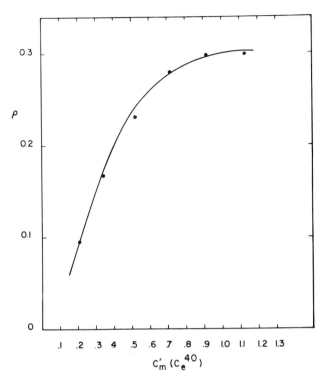

Fig. 25. The fraction of gelatin incorporated into the coacervate phase, following removal of the high molecular weight components by prior complex coacervation. Measurement at 30°, initial concentration at 40°.

of the gelatins in the two new phases were 1.67×10^5 and 1.87×10^5, showing that no further fractionation had occurred and that the system now represented a homogeneous coacervation situation. In this case, V_C was a nearly linear function of C'_M (the "new" mixing concentration, $C_E^{40°}$), C_C was constant within experimental precision, and ρ had the nearly ideal form (Fig. 25). However, the effect of mixing concentration was still evident since ρ did not approach a value of 1 and C'_E was still an increasing function of C'_M. The initial mixing concentration thus remained as a function regulating the coacervation equilibrium in the initially heterodisperse system.

D. Complexing in Systems with Deviations from Random Chain Character

A number of studies of electrostatic complexing phenomena have been examined under various guises and, although coprecipitation is a more usual consequence than coacervation, the pH, ionic strength, charge density, and mixing ratio dependencies are all similar to those described above.

Serum albumin has been shown to form complexes with heparin and other anionic polysaccharides but with varying intensity due to differences in the polysaccharide charge densities (40). Goldwasser and Putnam (41) used electrophoretic techniques to examine serum albumin–DNA complexes. They found that electrostatic complexes were formed even at pH values slightly basic to the albumin isoelectric pH. Coprecipitation occurred at pH's acid to the isoelectric point, but soluble complexes of varying composition were formed at higher pH. This type of behavior was more clearly illustrated by Rice et al. (42) who studied the interaction between bovine serum albumin and polylysine (degree of polymerization ~100). Both soluble and insoluble complexes could be formed, depending upon the ionic strength and pH. Under the strongest complexing conditions about 7 albumin molecules were complexed with a single polylysine molecule. Away from the equivalence point, excess polylysine and excess serum albumin both decrease the amount of precipitate. When polylysine is in great excess the soluble components do not appear to have constant composition but contain excess polylysine as judged from electrophoretic analyses. Morawetz and Hughes (43) examined the interaction of serum albumin with a variety of synthetic polyacids and polybases of high charge density. In most cases the complexes were quite resistant to disruption by the addition of salts, and only the mixture of serum albumin with methacrylic acid-diethylaminoethyl methacrylate copolymer, which is a polyampholyte, was sensitive to the addition of salts at low concentrations. At the maximum precipitation conditions, there were about 20 anionic charges on the polymer per serum albumin molecule in the complex. Bovine serum albumin has a pI ~5.4 and another protein, catalase, a pI ~5.0. Only a very small amount of polymer is required to precipitate catalase when the polymer is mixed with polyacrylic acid, while a larger amount is required to precipitate albumin under the same conditions (44). Catalase has a molecular weight several times that of serum albumin. However, when the albumin and catalase were mixed at the same time, only a single precipitate was observed and both catalase and albumin precipitate under the same conditions, similar to albumin alone. One could not, in this case, use the coprecipitation of polyion and protein to separate albumin from catalase.

Serum albumin also forms soluble complexes with lysozyme, a basic globular protein (pI = 10.9), at pH's between their isoelectric points (45). Steiner (45) showed that the complexes represent a continuous molecular weight distribution, that the association was reversible, and that

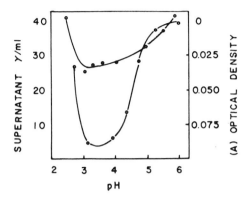

Fig. 26. Lower curve: Turbidities of solutions of RNA and RNAse as a function of pH (21 μg/ml RNA, 300 μg/ml RNAse). Upper curve: The solubility of RNA in solutions of RNAse as a function of pH (42 μg/ml RNA, 600 μg/ml RNAse) (*48*).

the electrostatic free energy of interaction was sufficiently large to account entirely for the complexing behavior. Lysozyme has a molecular weight of about 14,000, so a series of complexes with the heavier (MW ~ 70,000) albumin is not unexpected.

More recently, attention has been turned to the interaction of nucleic acids with basic proteins and polypeptides, primarily because of the possible role of proteins as regulatory substances in inducible enzyme synthesis (*46*) and the fact that histones apparently stabilize DNA in the conjugated nucleoproteins of cellular structures (*47*). Ribonuclease (RNAse) combines with RNA at pH < 4.5 by electrostatic interactions, and with DNA in the pH range 3–7 by nonspecific, electrostatic interactions (*48*). Figures 26 and 27 show the pH dependence of the RNAse

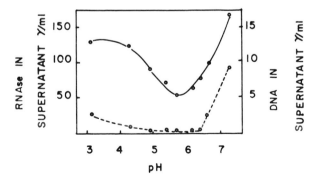

Fig. 27. The solubilities of DNA (lower curve) and RNAse (upper curve) as a function of pH (14 μg/ml DNA, 160 μg/ml RNAse) (*48*).

interaction with both RNA and DNA. These data of Coleman and Edel-hoch are very similar to the pH dependencies noted by Morawetz and Hughes (43) for serum albumin, polyion complexes. The DNA–RNAse interaction is reversible. RNA added to a DNA–RNAse complex will strip the RNAse from the complex and form a new RNA–RNAse complex. Felsenfeld et al. (49) obtained similar results and found, in addition, that RNAse destabilized DNA in terms of its susceptibility to thermal denaturation at $T < 60°$ but stabilized the structure at $T > 60°$. RNAse was preferentially bound to the denatured form of DNA. Further studies of DNA with basic polypeptides (50,51) showed the complexes formed to approach 1:1 electrical equivalence and that precipitation was greatest at this ratio (50). Because of the high charge densities, the complexes were very resistant to salts in some cases, resisting dissociation even at 2.0 M NaCl concentrations (51). The use of different basic polypeptides, such as poly-L-ornithine, poly-L-lysine, poly-L-arginine, and poly-L-homoarginine, showed that specific effects were also important in these interactions and particularly in the effect on stabilizing DNA against thermal denaturation. Similar results have been obtained by Liquori et al. (52) on the DNA–polyamine system. In this case a specific structural relationship has been proposed, based on crystallographic evidence, for the spermine–DNA system complex.

In most of these studies, which are of great interest and biological significance, general thermodynamic interpretations of the complexing reactions are very difficult because of the wide disparities in molecular weights and structures of the interacting species, leading to variations in the complexing stoichiometry, and because of specific, nonelectrostatic interactions frequently encountered. The thermodynamic arguments presented in the next section deal only with the most general case of random chain polyion interactions. It is hoped that such treatments can be applied eventually to these specific problems by the inclusion of appropriate specific structure and interaction effects.

III. THERMODYNAMICS OF THE DEMIXING PROCESS

In describing the experimental data in the preceding section it was convenient to separate electrostatic interaction terms from those apparently more closely related to chain configuration, molecular weight, and so on, because of the different nature of the experiments involved. It is not quite so easy to justify this separation when developing quantitative expressions to describe the phase equilibria. The first consequence upon mixing oppositely charged polymeric polyions, amply documented in Section II, is the formation of complexes or aggregates which form most favorably at mixing ratio appropriate for electrical equivalence. This aggregate formation is essentially electrostatic in nature, as indicated by the extreme sensitivity of the aggregates to microion salts at low polyion charge densities. Aggregate formation brings about a marked change in the system entropy, however, presumably leading to an unfavorable reduction in the entropy. When the electrostatic interaction is very great, as at high polyion charge densities, then coprecipitation occurs. However, at low charge density, and where the polyions are random-chain molecules, it appears that a gain in entropy may be possible by mixing the aggregated pairs into a single concentrated phase, the coacervate. (Molecules with fixed configurations will have a much lower tendency to form coacervate.) This demixing may occur without much sacrifice of the favorable electrostatic interaction in the aggregates because the disposition of charged groups in the coacervate must still be such as to minimize the electrostatic free energy as much as possible. Coacervate concentrations are usually quite low, on the order of 4% in the gelatin systems, and thus, on the basis of the general considerations of Section I.D and Fig. 2, the electrostatic interaction and entropy terms in the net free energy of mixing expression must have nearly equal magnitudes at low concentrations of polyion but the electrostatic contribution, $\Delta G_{M,i}$, must reach a limiting or asymptotic value, less than the entropy contribution, ΔG_M^a at a quite low value of the concentration. Salt suppression is a direct consequence of this model, since increasing ionic strength should reduce $\Delta G_{M,i}$.

Self-suppression, molecular weight fractionation, and selection in heterogeneous polyion mixtures can also be thought of in qualitative terms as a consequence of the electrostatic interactions. Consider the concentrated phase first. Assuming that both polyions, P^+ and Q^-, have the same charge densities, $\pm\sigma$, per unit weight or segment, then at electrical equivalence the distribution of charges throughout the coacervate phase will be essentially independent of the molecular weights of the components distributed in P^+ and Q^-. In fact, as we shall demonstrate later, the entropy gain on transfer of a given weight of small polymer from aggregate to concentrated phase is more favorable than the transfer of an equal weight of larger polymer and should lead to the accumulation of lower weight components in the coacervate! The situation is quite different in the dilute phase. Suppose P^+ and Q^- have equal molecular weights and charge densities. Pair-wise aggregates will then be neutral and $\Delta G_{M,i}$ minimized. If P^+ and Q^- differ in weight but have similar charge density, pair-wise aggregates would bear excess charges, or larger aggregates of less favorable $\Delta G_{M,i}$ would be formed. For example, if at pH 7, gelatin P^+ had a weight of 1×10^{-5} and Q^- 1.5×10^5 but each had $\sigma = 10$ eq/

10^5 g, then the minimum size neutral aggregate would contain 5 molecules, 3 of P^+ and 2 of Q^-. Electrophoretic studies (*15,41,42*) suggest that under such circumstances the "soluble complexes," i.e., the dilute phase aggregates, depend for their composition on mixing ratio and do bear excess net charges. Thus, unequal weights, weight distributions, or charge densities will lead to aggregates with values of $\Delta G_{M,i}$ less favorable for phase separation than in the completely symmetric case. If the molecular weight distributions of P^+ and Q^- overlap, minimization of $\Delta G_{M,i}$ on mixing and aggregate formation demands pairing of comparable weight components of each polyion.

Upon preparing a symmetric low concentration mixture of P^+ and Q^- of similar molecular weight distribution the solution then becomes a mixture of neutral PQ aggregates of different molecular weight, each of which may be considered to be new components in a multi-component, solvent, and $(PQ)_i$ mixture. In addition to the proposal indicated above that phase separation from such a mixture might result from the entropy increase on mixing aggregates in the concentrated phase, one may also consider that phase separation takes place due to the incompatibility (i.e., positive χ_{23}) of the aggregate components as discusssed in Section I.C, leading to the appearance of discrete coacervate phases.

The following discussions present the attempts which have been made to obtain quantitative expressions for these qualitative ideas.

A. The Voorn–Overbeek Random Phase Model

Voorn (*13,14,24,53,55*) and Overbeek and Voorn (*56,57*) developed a theory for coacervation based upon the following assumptions.

(1) Each polyion in each phase had a random chain configuration whose entropy was given by the Flory–Huggins equation, Eq. 4.

(2) Solvent–solute interactions were negligible compared with electrostatic interactions, i.e., set $\chi_{12} = 0$.

(3) The polyions in both phases may be treated as the sum of single charges, i.e., they are thought of as distributed throughout the solution as if they were independent of the chains on which they reside.

With these assumptions the free energy of mixing polymer with solvent in both phases is given by Eq. 5 where $\Delta G_{M,i}$ is taken as entirely the result of electrostatic interactions, $\Delta G_{M,i} = \Delta G_{elect}$. Voorn (*53*) used the Debye–Hückel treatment to evaluate ΔG_{elect}. Consider a volume of solution containing M_0 lattice sites, with each solvent molecule of volume v occupying one lattice site, $r_1 = 1$. Each polymer molecule will occupy M lattice sites, $r_2 = M$, and will bear a total charge $M\sigma$. The charge density, σ, thus indicates the average fractional charge per polymer lattice site and the total charge per polyion is $M\sigma$. From the Debye–Hückel theory,

$$\frac{\Delta G_{\text{elect}}}{kT} = \frac{-K^3 v}{12\pi} \tag{34}$$

where K is the reciprocal of the ionic atmosphere or

$$K^2 = \frac{4\pi e^2}{D_0 kT} \sum_i \rho_i z_i^2 \tag{35}$$

ρ_i = number density of point charges of type i.
z_i = number of electrons on point charge i.
e = charge on electron.
D_0 = dielectric constant of the medium.

Then, if φ_2 is the total volume fraction of polymer in the symmetric P^+,Q^- mixture

$$\rho_i = \frac{\varphi_2 \sigma}{2v}$$

And for monovalent point charges

$$K^2 = \frac{4\pi e^2}{D_0 kT} \varphi_2 \sigma \tag{36}$$

Equation 5 can then be written

$$\frac{\Delta G_M}{M_0 kT} = \left[\varphi_1 \ln \varphi_1 + \frac{\varphi_2}{M} \ln \varphi_2 - Y\varphi_2^{3/2}\sigma^{3/2} \right] \tag{37}$$

where

$$Y = \left(\frac{4\pi}{9v} \right)^{1/2} \left(\frac{e^2}{D_0 kT} \right)^{3/2} \tag{38}$$

Y is a dimensionless quantity which depends on the solvent and temperature but is independent of the nature of the polyions. If one wishes to include solvent–solute interactions, Eq. 37 is readily modified by the inclusion of the χ_{12} parameter to yield

$$\frac{\Delta G_M}{M_0 kT} = \left[(1 - \varphi_2) \ln (1 - \varphi_2) + \chi_{12}(1 - \varphi_2)\varphi_2 + \frac{\varphi_2}{M} \ln \varphi_2 - Y\varphi_2^{3/2}\sigma^{3/2} \right] \tag{39}$$

The chemical potentials,

$$\Delta\mu_1 = \left(\frac{\partial \Delta G_M}{\partial N_1} \right)_{T,N_2}, \qquad \Delta\mu_2 = \left(\frac{\partial \Delta G_M}{\partial N_2} \right)_{T,N_1}$$

follow directly from Eq. 39.

$$\frac{\Delta\mu_1}{kT} = \ln(1 - \varphi_2) - \varphi_2\left(1 - \frac{1}{M}\right) + \chi_{12}\varphi_2^2 + \frac{Y\varphi_2^{3/2}\sigma^{3/2}}{2} \qquad (40)$$

$$\frac{\Delta\mu_2}{kT} = \ln\varphi_2 - (M - 1)(1 - \varphi_2) + \chi_{12}M(1 - \varphi_2)^2 - \frac{Y\sigma^{3/2}}{2}\varphi_2^{1/2}(3 - \varphi_2) \qquad (41)$$

The critical conditions for phase separation can be deduced by application of the conditions of Eq. 9 to Eq. 40 or those of Eq. 10 to Eq. 41. The critical conditions, using Eqs. 9 and 40, are thus

$$\frac{\partial\left(\frac{\Delta\mu_1}{kT}\right)}{\partial\varphi_2} = 0 = -\frac{1}{1 - \varphi_2} + 1 - \left(\frac{1}{M_C}\right) + 2\chi_C\varphi_2 + 3/4Y\sigma^{3/2}\varphi_2^{1/2} \qquad (42)$$

and

$$\frac{\partial^2\left(\frac{\Delta\mu_2}{kT}\right)}{\partial\varphi_2^2} = 0 = -\frac{1}{(1 - \varphi_2)^2} + 2\chi_C + 3/8Y\sigma^{3/2}\varphi_2^{-1/2} \qquad (43)$$

The symbols χ_C and M_C refer to the critical values of these parameters. Combining Eqs. 42 and 43, one finds

$$2\chi_C = \frac{1}{(1 - \varphi_2)^2} - 3/8Y\sigma^{3/2}\varphi_2^{-1/2} \qquad (44)$$

and

$$\frac{1}{M_C} = \frac{\varphi_2^2}{(1 - \varphi_2)^2} + 3/8Y\sigma^{3/2}\varphi_2^{1/2} \qquad (45)$$

When $\chi_C = 0$, as required by the Voorn restriction described earlier, then

$$M_C = \frac{(1 - \varphi_2)^2}{\varphi_2(1 + \varphi_2)} \qquad (46)$$

and

$$\sigma_C = \frac{64}{9Y^2}\frac{\varphi_2}{(1 - \varphi_2)^4} \qquad (47)$$

Combining Eqs. 46 and 47, and assuming that φ_2 is a small number

$$M_C\sigma_C^3 = \frac{64}{9Y^2} \qquad (48)$$

For $T = 298°$, $v = 3 \times 10^{-23}$ cm³, and $D_0 = 85$, Y has the value 3.665 and, hence, $M_C\sigma^3 = 0.53$. Thus, on the basis of the random-chain distribution in both phases, a zero value for χ_{12}, and uniform charge distributions, complex coacervation will occur only if $M_C\sigma^3 \geq 0.53$. In the case of the isoionic gelatin mixtures examined in detail, reasonable values for the high molecular weight fractions A-I and B-I at pH 7 are $M \approx 10^5$ and $\sigma \approx 2 \times 10^{-3}$, or $M\sigma^3 \approx 8 \times 10^{-4}$, orders of magnitude smaller than required to predict a phase separation by the Voorn-Overbeek theory with $\chi = 0$. σ is too low by at least a factor of 10.

To determine the validity of the form of the electrostatic term and the effect of neglecting χ_{12}, Veis (37) modified the Voorn–Overbeek treatment to include χ_{12} to yield the chemical potential equations given in Eqs. 40 and 41. At phase equilibrium, the solvent chemical potential expression yields

$$\ln \frac{\varphi_1^I}{\varphi_1^{II}} - (\varphi_1^I - \varphi_1^{II}) = 4\chi_{12}[(\varphi_2^{II}) - (\varphi_2^I)^2] + Y'\sigma^{3/2}[(\varphi_2^{II})^{3/2} - (\varphi_2^I)^{3/2}] \quad (49)$$

The volume of fractions and charge density were available from experiment so that χ_{12} could be evaluated. When applied to data on heterogeneous systems, such as that exemplified by Figs. 14 and 15, use of the electrostatic term with a $3/2$ power dependence on the phase concentrations gave meaningless and inconsistent results for χ_{12}. An alternate solution was to take the electrostatic term, which is related to difference between electrostatic free energies in each phase, i.e., $\Delta(\Delta G_{\text{elect}})/kT$, as a parameter of the system. Hence,

$$\ln \frac{\varphi_1^I}{\varphi_1^{II}} - (\varphi_1^I - \varphi_1^{II}) = 4\chi_{12}[(\varphi_2^{II})^2 - (\varphi_2^I)^2] + f_1(\sigma, \varphi) \quad (50)$$

and taking sets of phase composition data in narrow regions of C_M, it was possible to compute compatible average values of χ_{12} and $f_1(\sigma,\varphi)$ as a function of C_M. It was immediately apparent from such computations, as illustrated in Fig. 28, that neither χ_{12} nor $f_1(\sigma,\varphi)$ was independent of C_M and that, at low C_M, the electrostatic free energy term was actually unfavorable for phase separation (37). At higher C_M where χ_{12} becomes small, $f_1(\sigma,\varphi)$ assumes favorable (positive) values for phase separation. In the region where self-suppression was evident, $C_M > 0.8\%$, $\chi_{12} \approx 0.5$, and $f_1(\sigma,\varphi) \approx 0$. Similar computations have been made for the systems of pauci-disperse gelatins where plateau values of C_E and C_C are approached, as in Fig. 18. Along any C_E plateau the values of χ_{12} and $f_1(\sigma,\varphi)$ are constant, but vary consistently from one plateau region to the next. Figure 29 shows the χ_{12}–$f_1(\sigma,\varphi)$ relationships for the data of Figs. 14 and 18. In spite of the wide divergences in the C_M dependence of ρ and V_C in these two cases, the dependencies of χ and $\Delta(\Delta G_{\text{elect}})$ on C_M are similar. Coacervation is most intense as one moves to the right in Fig. 29 (to lower C_M). As defined in Eq. 50, $f_1(\sigma,\varphi)$ is $(\Delta\mu_{1,\text{elect}}^{II} - \Delta\mu_{1,\text{elect}}^I)$

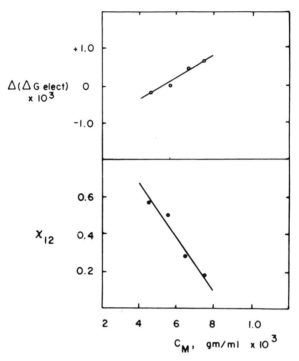

Fig. 28. The solvent–solute and electrostatic free energy parameters, computed according to Eq. 50, for unfractionated A + B gelatin mixtures at 40° (*34*).

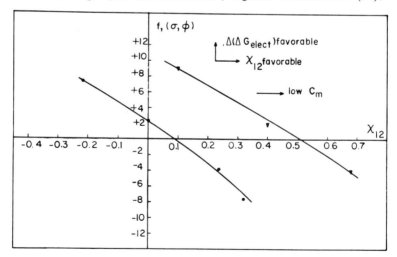

Fig. 29. The relationship between χ_{12} and $\Delta(\Delta G_{elect})$ in gelatin–gelatin coacervation. (▲) Heterogeneous gelatin (refer to Fig. 14). (●) Pauci-disperse gelatin (refer to Fig. 18).

$$RT = -\frac{M_2}{M_1}\frac{(\Delta\mu_2^{II} - \Delta\mu_2^{I})}{RT}$$

and represents the negative of the electrostatic free energy change on trans-fer of a polymer segment from the dilute to the concentrated phase.   Posi-tive values of $f_1(\sigma,\varphi)$ therefore favor coacervation.   Thus, the fact that $f_1(\sigma,\varphi)$ is negative in the lower right-hand quadrant of Fig. 29, where coacervation is most intense, indicates that the electrostatic free energy contribution is unfavorable for phase separation and the process is driven by the positive χ_{12} term.

These considerations lead to the inescapable conclusion that, although the pH and salt dependence of coacervation clearly indicate an important and crucial role in coacervation for electrostatic interactions, the basic postulate of Voorn suggesting that the electrostatic contribution drives the demixing process cannot be valid.   The key element in the Voorn analysis is that of treating the charges as if they were independent of the chains on which they reside and were distributed uniformly in both phases.   As we have seen, this is an untenable assumption both from the phase equilibrium equations and the direct experimental observation of aggregates in the dilute phase.

B. The Symmetrical Aggregate Model

The approach taken by Veis (*34,37*), Veis and Bodor (*23*), and Veis et al. (*33*) was to treat the symmetrical mixing system in terms of differ-ent standard states in the two phases.   In the absence of any data indicat-ing ordering in the concentrated phase, the Voorn model of randomly mixed polymers was taken for the coacervate phase but the dilute phase was considered to be composed of symmetrical aggregates.   The first step upon mixing was assumed to be an immediate pair-wise aggregation be-tween oppositely charge polyions:

$$P^+OH^- + H^+Q^- \rightarrow (PQ)_{agg} + H_2O$$

The entropy change for this reaction was unfavorable while the neutraliza-tion of charges made the aggregation very favorable, thus $\Delta S_{agg} < 0$, $\Delta G_{elect} << 0$.   The PQ aggregates thus became the new solute compo-nent for the mixture at C_M.   Demixing then involved the aggregates:

$$(PQ)_{agg,C_M} \rightarrow [PQ_{agg}]^I + [P^+ + Q^-]_{random}^{II}$$

In this case it is assumed that $\Delta S > 0$ for the transfer of polymer from aggregate to random state in the concentrated phase, while $\Delta G_{elect} \geq 0$, somewhat unfavorable to the phase separation.

1. The Chemical Potentials of Solvent and Solute in the Dilute Aggregate Phase

The chemical potential of solvent in a dilute polymer solution can be written in terms of the excluded volume, U, as

$$\frac{\mu_1 - \mu_1^0}{kT} = -\left[n_2 + \frac{U_2}{2} n_2^2\right]\left(\frac{\partial V}{\partial N_1}\right)_{E,V,N_2} \tag{51}$$

or in terms of a virial equation such as

$$\frac{\mu_1 - \mu_1^0}{kT} = -[n_2 + B_2 n_2^2]v \tag{52}$$

In these equations n_2 is the number density of polymer molecules, N_1 the mole fraction of solvent, v a constant $= (\partial V_1/\partial N_1)_{N_2}$, and B_2 is the second virial coefficient in which one includes all deviations from ideal behavior. Flory (4) and Orofino and Flory (58,59), showed that for random chain polymers

$$B_2 = \frac{U_2}{2} = \frac{2^{5/2}\pi \overline{r^2}^{3/2}}{27} - \ln\left[1 + \frac{\pi^{1/2}}{2}(\alpha^2 - 1)\right] \tag{53}$$

where $\overline{r^2}$ is the mean square end-to-end extension of a polymer molecule and α is a distortion parameter which expresses the deviation of the polymer dimensions from that computed for the unperturbed situation in which the excluded volume is zero. If the end-to-end extension is increased by the factor α due to overlaps of chain segments, each segment can be taken as having the same average increased projection in the direction of the end-to-end vector. When $\alpha = 1$, $B_2 = 0$. Equation 53 is based on the use of a Gaussian distribution of polymer segments through the polymer molecule domain with maximum segment density at the center. This equation is an approximation valid where $\alpha \geq 1$, but not much greater than 1. When $\alpha \leq 1$, B_2 is negative and the approximation for B_2 is better represented (4) by

$$B_2 = \frac{2^{3/2}\pi^{3/2}\overline{r^2}^{3/2}}{27}(\alpha^2 - 1) \tag{54}$$

The solvent chemical potential is evidently quite sensitive to the value of the distortion parameter, α. The symmetrical aggregate "component" consists of two oppositely charged molecules which must have their centers of gravity located together within some small volume element, δV, if the individual molecules have flexible chain conformations. If each chain is presumed to bear charges of only one kind, then in order to minimize the electrostatic free energy of the aggregates it is likely that the two molecules will have the same average end-to-end extension at all times or, in other words, their conformational fluctuations and end-to-end distances will be highly correlated. As an approximation to this situation, and baring specific charge group pairings, the aggregate can be treated as if it consists of two equal length chains crosslinked at their centers but with the charges distributed independently of their residence on the chains. Under these circumstances, Gates (60) has shown that

$$B_{2,\text{agg}} = \frac{2^{5/2}\pi \overline{r^2}^{3/2}}{27} \ln[1 + \pi^{1/2}(\alpha^2 - 1)]; \qquad \alpha \geq 1 \tag{55}$$

and

$$B_{2,\text{agg}} = \frac{2^{5/2}\pi \overline{r^2}^{3/2}}{27} (\alpha^2 - 1); \qquad \alpha \leq 1 \tag{56}$$

Each expression differs only by a factor of two from the corresponding Eqs. 53 and 54.

The distortion of the aggregate can be determined by calculating (60) the total free energy within the domain of an aggregate and minimizing this free energy with respect to α. We assume that within the aggregate the electrostatic contribution at low charge density can be represented by a Debye–Hückel term, an obvious oversimplification and perhaps a poor approximation. The free energy of the aggregate is given by

$$\frac{\Delta G}{kT} = \sum_i \frac{\delta(\Delta G_{\text{mix}})_i}{kT} + \frac{\Delta G_{\text{distortion}}}{kT} \tag{57}$$

$\delta(\Delta G_{\text{mix}})_i$ is the free energy of mixing polymer segments and solvent within the volume element δV_i within the domain of an individual aggregate. Assuming that the interior of the aggregate will be much like the polymer in concentrated solution, but with M set equal to infinity because there is only a single molecule and no polymer mixing, Eq. 39 can be used directly in the form

$$\frac{\delta(\Delta G_{\text{mix}})_i}{kT} = \frac{\delta V_i}{v}[(1 - \varphi_{2i}) \ln (1 - \varphi_{2i}) + (1 - \varphi_{2i})\varphi_{2i} - Y\sigma^{3/2}\varphi_{2i}^{3/2}] \tag{58}$$

In Eq. 58, φ_{2i} refers to the volume fraction of polymer within the polymer aggregate domain and is not the bulk average concentration of polymer, φ_2^I. Wall (61,62) has shown that the entropy of distortion of a single polymer molecule composed of freely jointed segments by an expansion of α is given by

$$\frac{\Delta S_{\text{distortion}}}{k} = \ln \alpha^3 - \tfrac{3}{2}(\alpha^2 - 1) \tag{59}$$

For two molecules with M segments each joined in the aggregate by a hypothetical crosslink at their midpoints, $\Delta S_{\text{distortion}}$ within the aggregate will be twice that given in Eq. 59. The contribution to the free energy will be just that of the entropy of distortion since the electrostatic energy term has been included in Eq. 58. Hence,

$$\frac{\Delta G_{\text{distortion}}}{kT} = 3(\alpha^2 - 1) - 2 \ln \alpha^3 \tag{60}$$

If we now take δV_i as the volume of a spherical shell located at a distance S_i^α from the center of gravity of the aggregate and having a thickness $\alpha \delta S_i$, then

$$\delta V_i = 4\pi \alpha^3 S_i^2 \delta S_i \tag{61}$$

Using this terminology, S_i refers to a distance from the center of gravity in an undistorted molecule. For a pair-wise aggregate, the total number of polymer segments in this spherical shell is given by a Gaussian distribution

$$2M_i = 2M \left(\frac{27}{\overline{r_0^2}^{3/2} \pi^{3/2}} \right) \exp \left(- \frac{9 S_i^2}{\overline{r_0^2}^{3/2}} \right) 4\pi S_i^2 \delta S_i \tag{62}$$

where $\overline{r_0^2}$ is the unperturbed mean square end-to-end extension $\overline{r_{\alpha=1}^2} = na^2$. These $2M_i$ segments occupy the shell between S_i and δS_i in the perturbed case. The number of solvent molecules in this shell is

$$N_{1i} = \frac{4\pi \alpha^3 S_i^2 \delta S_i (1 - \varphi_{2i})}{v} \tag{63}$$

The volume fraction occupied by polymer segments within the aggregate is

$$\varphi_{2i} = \frac{2M_i v}{\delta V_i} = 2Mv \left(\frac{27}{\overline{r_0^2}^{3/2} \pi^{3/2} \alpha^3} \right) \exp \left(\frac{9}{\overline{r_0^2}^{3/2}} S_i^2 \right) \tag{64}$$

The free energy of the aggregate is minimized with respect to α when

$$\sum_i \frac{\partial [\delta (\Delta G_{\text{mix}})_i]}{\partial \alpha} + \frac{\partial \Delta G_{\text{dist}}}{\partial \alpha} = 0 \tag{65}$$

Utilizing Eqs. 58 and 60, Eq. 65 may be reduced to

$$6\alpha - \frac{6}{\alpha} = \sum_i \left[\left(\frac{1}{2} - x \right) \varphi_{2i}^2 + \frac{\varphi_{2i}^3}{3} - \frac{Y \sigma^{3/2} \varphi_{2i}^{3/2}}{2} \right] \frac{\partial N_{1i}}{\partial \alpha} \tag{66}$$

From Eqs. 62 and 63,

$$\frac{\partial N_{1i}}{\partial \alpha} = \frac{12\pi \alpha^2 S_i^2 \delta S_i}{v} \tag{67}$$

and upon replacing the summation of Eq. 66 by an integral taken over all S_i, one obtains

$$(\alpha^2 - 1) = \left(\frac{1}{2} - \chi\right) M\theta + \left(\frac{2^3}{3^{5/2}}\right) M\theta^2 - \left(\frac{2^{5/4}}{3^{3/2}}\right) Y\sigma^{3/2} M\theta^{1/2} \qquad (68)$$

where

$$\theta = \frac{27(2Mv)}{(2\pi\overline{r^2})^{3/2}} \qquad (69)$$

Orofino and Flory (58), in a comparable treatment of the excluded volume problem for single polymers, defined θ as the effective average or uniform volume fraction of polymer segments within the polymer domain. If we take the aggregate domain as a sphere with radius R, then

$$\theta = \frac{2Mv}{4/3\pi R^3} \qquad (70)$$

and, comparing Eqs. 69 and 70, $2R \approx 1.04\overline{r^2}^{1/2}$.   Since $\overline{r^2} = \alpha^2 r_0^2$, then we can define a θ_0 such that

$$\theta_0 = \alpha^3 \theta \qquad (71)$$

and Eq. 68 can be rewritten as

$$\frac{\alpha^3(\alpha^2 - 1)}{M\theta_0} = \frac{1}{2} - \chi + 0.5132 \frac{\theta_0}{\alpha^3} - 0.4577 Y\sigma^{3/2}\left(\frac{\alpha^3}{\theta_0}\right)^{1/2} \qquad (72)$$

For larger positive values of χ and at high charge densities, the first and third terms on the right-hand side of Eq. 72 predominate and $\alpha < 1$. The approximations involved in arriving at Eq. 72, particularly in obtaining Eq. 68, restrict this equation to small values of θ_0, on the order of ≤ 0.1, but this is entirely within the range of experimental observation. The concentrated phase in a typical coacervation system has $\varphi_2^{II} \leq 0.05$, and these solutions must have densities greater than that of the aggregates with which they are in equilibrium.

From the expressions for the second virial coefficient, Eqs. 55 and 56, substitution of Eqs. 68, 69, and 70 yield

$$B_{2,\text{agg}} = \frac{(2M)^2 v\alpha^3}{\theta_0 M \pi^{1/2}} \ln\left[1 + \pi^{1/2}(\alpha^2 - 1)\right]; \qquad \alpha \geq 1 \qquad (73)$$

and

$$B_{2,\text{agg}} = \frac{(2M)^2 v\alpha^3(\alpha^2 - 1)}{\theta_0 M} = (2M)^2 v\left[\frac{1}{2} - \chi + 0.5132\left(\frac{\theta_0}{\alpha^3}\right)\right.$$
$$\left. - 0.4577 Y\sigma^{3/2}\left(\frac{\alpha^3}{\theta_0}\right)^{1/2}\right]; \qquad \alpha \leq 1 \qquad (74)$$

It is interesting to note that Flory (4) used the approximation $B_2 \sim M^2v(\frac{1}{2} - \chi)$ in treating phase separations in systems of uncharged single polymer molecules. This corresponds to dropping the pairing restriction on the aggregate and eliminating the electrostatic and higher-order virial expansion terms in Eq. 66. Orofino and Flory (58,59) concluded that the term $+0.5132(\theta_0/\alpha^3)$ would be negligible. However, when χ exceeds 0.5, α is less than 1 while $\frac{1}{2} - \chi$ may be close to 0. Thus, this approximation is questionable for highly interacting systems.

The chemical potential of solvent in the dilute phase of aggregates (Eq. 52) written in terms of volume fractions and with reference state of pure solvent is

$$\frac{\Delta\mu_1}{kT} = - \frac{\varphi_2}{2M} - \frac{B_2}{(2M)^2v} \varphi_2^2 \tag{75}$$

where φ_2 once again refers to the average volume fraction occupied by the aggregates in the solution so that $\varphi_2 \ll \theta_0$. From the Gibbs–Duhem relationship one finds

$$\frac{\Delta\mu_2}{kT} = \frac{1}{2}\ln\varphi_2 + \frac{1}{2}(1 - \varphi_2) - \frac{MB_2}{(2M)^2v}(1 - \varphi_2)^2 \tag{76}$$

The reference state for the solute is a pure, disordered collection of polymer chains in paired aggregates. The total free energy of the dilute phase is thus

$$\frac{\Delta G}{kT} = n_1\frac{\Delta\mu_1}{kT} + n_2\frac{\Delta\mu_2}{kT} = M_0\left[\frac{\varphi_2}{2M}\ln\varphi_2 - \frac{B_2}{(2M)^2v}\varphi_2(1 - \varphi_2)\right] \tag{77}$$

2. The Chemical Potentials in the Concentrated Phase

Assuming the random distributions of polymer segments and charges as in the Voorn treatment, Eqs. 40, 41, and 39, for the concentrated phase, the appropriate counterparts are Eqs. 75, 76, and 77, respectively. Comparing any pair of these, e.g., Eqs. 39 and 77 for the total free energy of each phase, it is apparent that the forms of the two equations are different. In addition, the ranges of validity of the two expressions differ. In order to have the polymer segments distributed uniformly and undistorted in the concentrated phase, as required by the model, the volume fraction of polymer in that phase must exceed the volume fraction occupied by polymer segments in the volume element δV in an aggregate when $\alpha = 1$. This maximum volume fraction within the aggregate occurs when $S_i = 0$, $\alpha = 1$. From Eq. 64 this is seen to be the condition where

$$\varphi_2^{II} \geq 2^{3/2}\theta_0 \tag{78}$$

and Eq. 39 is valid only in this range. On the other hand, Eq. 77 will be valid only when $\varphi_2^{\mathrm{I}} \ll \theta_0$ and the proper form of $B_{2,\mathrm{agg}}$ will be that where $\alpha \leq 1$, Eq. 74.

3. The Standard State Entropy Change in the Dilute Phase

In order to equate the chemical potentials in each phase and deduce the phase compositions, it is necessary to have both sets of chemical potentials referenced to the same standard state. As described above, the standard state for the neutral aggregates is pure, disordered random chain polymer in pairwise aggregates whereas the standard state for the concentrated phase is that of pure, individual random chain molecules. For the symmetric mixing case, the rectification of standard states is accounted for by the process

$$\frac{N_2}{2} \text{ (aggregates of } 2M)_{\varphi_{2-1}} \to N_2 \text{(molecules of } M)_{\varphi_{2-1}} \qquad (79)$$

for which $\Delta S > 0$. N_2 is the number of individual polymer molecules. At $\varphi_{2=1}$ we assume that there will be no energy change on "unpairing" so that only combinatorial factors will be of interest. Hence, only the entropy of the process need be considered. Gates (60) has approximated this entropy change as

$$\frac{\Delta S}{k} = \frac{N_2}{2} \ln \left(\frac{2V}{\delta V}\right) \qquad (80)$$

where V is the total volume of solution in which the N_2 polymer molecules are placed and δV is the size of the volume elements in which the centers of gravity of both polymers of each pair are located. Since each polymer segment occupies a lattice site volume v, the minimum value for δV has been taken as $2v$. Consequently,

$$\frac{\Delta S}{k} = \frac{N_2}{2} \ln N_2 M \qquad (81)$$

This equation represents a drastic underestimate of ΔS since the calculation assumes random placement of the polymer segments within the aggregate. Obviously, one could not have regions with excess uncompensated net charge with any very great probability and, as stated earlier, it is likely that there is some correlation of end-to-end extensions. Nevertheless, Eq. 81 represents a reasonable first approximation and shows very clearly, in the form of Eq. 82, that for a given total volume fraction

$$\frac{\Delta G_{\mathrm{std\ state}}}{kT} = -M_0 \left[\frac{\varphi_2}{2M} \ln N_2 M\right] \qquad (82)$$

of polymer, the standard state change becomes smaller and smaller as M

increases so that, entropywise, aggregates between high molecular weight components are most easily formed.

The total free energy of the dilute phase with random, single chains as standard state is then

$$\frac{\Delta G}{kT} = M_0 \left[\frac{\varphi_2}{M} \ln \varphi_2 + \frac{\varphi_2}{2M} \ln \frac{V}{v} - \frac{B_2}{(2M)^2 v} \varphi_2 (1 - \varphi_2) \right] \quad (83)$$

and the corresponding chemical potentials are

$$\frac{\Delta \mu_1}{kT} = -\frac{\varphi_2}{2M} - \frac{B_2}{(2M)^2 v} \varphi_2^2 \quad (84)$$

and

$$\frac{\Delta \mu_2}{kT} = \frac{1}{2} \ln (\varphi_2^2 M_0) + \frac{1}{2} (1 - \varphi_2) - \frac{M B_2}{(2M)^2 v} (1 - \varphi)^2 \quad (85)$$

Equation 84 is identical with Eq. 75 for the chemical potential of the solvent taken without regard to the standard state of the solute. Hence, by working with the solvent equilibria the results will become independent of the nature of the assumptions concerning the aggregates, except for the explicit form of the second virial coefficient.

4. The Conditions for Phase Separation

According to the assumptions made concerning the nature of the aggregate and the appropriate ways of writing the chemical potentials, the pertinent expressions for comparing theory with the experimental data are Eqs. 84 and 40.

$$\frac{\Delta \mu_1^{II}}{kT} = \ln (1 - \varphi_2^{II}) + \varphi_2^{II} \left(1 - \frac{1}{M} \right) + \chi (\varphi_2^{II})^2 + \frac{Y \sigma^{3/2}}{2} (\varphi_2^{II})^{3/2}, \quad (40)$$

when $\varphi_2^{II} \geq 2^{3/2} \theta_0$ and

$$\frac{\Delta \mu_1^{I}}{kT} = -\frac{\varphi_2^{I}}{M} - \frac{B_2}{(2M)^2 v} (\varphi_2^{I})^2 \quad (84)$$

when $\varphi_2^{I} \ll 2^{3/2} \theta_0 < \varphi_2^{II}$ and

$$B_2 = (2M)^2 v \left[\frac{1}{2} - \chi + 0.5132 \left(\frac{\theta_0}{\alpha^3} \right) - 0.4577 Y \sigma^{3/2} \left(\frac{\alpha^3}{\theta_0} \right)^{1/2} \right]$$

when $\alpha \leq 1$. As one test of the adequacy of these expressions, and to assess the relative importance of the various terms, the experimental data of Fig. 18 may be used, just as in the analysis of the Voorn equations. Setting the chemical potentials in each phase equal and combining terms, one obtains

$$\ln(1 - \varphi_2^{II}) + \varphi_2^{II} - \frac{1}{M}(\varphi_2^{II} - \varphi_2^{I}) + \chi[(\varphi_2^{II})^2 - (\varphi_2^{I})^2]$$

$$+ \left[0.5 - 0.46 Y \sigma^{3/2} \left(\frac{\alpha^3}{\theta_0}\right)^{1/2} + 0.5\left(\frac{\theta_0}{\alpha^3}\right)\right](\varphi_2^{I})^2 + 0.5 Y \sigma^{3/2}(\varphi_2^{II})^{3/2} = 0$$

$$(86)$$

Table IV lists values for these various terms for the three concentration plateau regions of Fig. 18. The parameters listed at the top of Table IV represent either the best estimates (M, σ) or reasonable maximum (θ_0) or minimum (α) values. The phase concentrations are taken directly from experiment. On a term-for-term basis, for the case where $C_M = 0.3\%$, Eq. 86 becomes

$$-0.030406 + 0.02990 - 2.83 \times 10^{-7} + 8.92 \times 10^{-4}\chi + 1.29 \times 10^{-6}$$
$$+ 3.169 \times 10^{-7} = 0$$

The difference between the first two terms is on the order of 10^{-4}, the term in M^{-1} becomes significant only when $M < 10^3$ and the second virial coefficient term, while larger than the electrostatic term relating to the concentrated phase, would be essentially negligible even if σ were increased by a factor of 10. The dominant interaction term is that containing χ_{12}. With the data of Table IV, χ_{12} has values of 0.63, 0.56, and 0.62 for

TABLE IV

Analysis of Eq. 86, Comparison of Theory and Experiment
for Pauci-Disperse Mixtures

Adjustable parameters: $M = 10^5$, $\sigma = 2 \times 10^{-3}$
Assumed values: $\theta_0 = 0.01$, $\alpha = 0.9$, $v = 3 \times 10^{-23}$ cm^3/lattice site
Constants: $Y = 3.69$, $T = 313°$

$C_M, \%$	0.3	0.8	1.2
1. φ_2^{II}	0.0299	0.0267	0.0255
2. φ_2^{I}	0.0016	0.0036	0.0073
3. $\ln(1 - \varphi_2^{II})$	-0.03406	-0.0271	-0.0259
4. $(\varphi_2^{II} - \varphi_2^{I})/M$	$0.0283\ M^{-1}$	$0.0231\ M^{-1}$	$0.0160\ M^{-1}$
5. $[(\varphi_2^{II})^2 - (\varphi_2^{I})^2]\chi$	$8.92 \times 10^{-4}\chi$	$6.99 \times 10^{-4}\chi$	$5.97 \times 10^{-4}\chi$
6. $-0.4577 Y \sigma^{3/2}\left(\frac{\alpha^3}{\theta_0}\right)^{1/2}$	-1.29×10^{-3}	-1.29×10^{-3}	-1.29×10^{-3}
7. $0.5\frac{\theta_0}{\alpha^3}$	6.86×10^{-3}	6.86×10^{-3}	6.86×10^{-3}
8. $\left[0.5 - 0.46 Y \sigma^{3/2}\left(\frac{\alpha^3}{\theta_0}\right)^{1/2}\right.$ $\left. + 0.5\left(\frac{\theta_0}{\alpha^3}\right)\right](\varphi_2^{II})^2$	1.29×10^{-6}	6.56×10^{-6}	26.9×10^{-6}
9. $0.5\sigma Y^{3/2}(\varphi_2^{II})^{3/2}$	9.5×10^{-7}	8.05×10^{-7}	7.46×10^{-7}

$C_M = 0.3$, 0.9, and 1.2, respectively. Considering the lack of precision in the fourth significant figure for the phase compositions, this is excellent agreement and we can take χ_{12} as a constant over the entire range, ~0.60. Continuing the analogy with the analysis of the Voorn equation, these data suggest that the basic hypothesis of the aggregate model is correct, i.e., there is no appreciable change (and, in fact, a slight decrease) in electrostatic free energy on transfer of a polymer segment from the dilute to the concentrated phase.

Direct estimates of the free energies are possible with the use of Eqs. 39 and 83, taking χ_{12} as 0.60. The numerical results of the computations are given in Table V. The dilute phase free energy, ΔG^I, is dominated by the interaction terms in B_2. The standard state change also operates to make the free energy more positive. The standard state change term becomes important as χ_{12} approaches its critical value. In every case, however, the combinatorial term is small compared with the standard state change term. In marked contrast, the combinatorial terms provide the major contribution to the free energy of the concentrated phase. The heat of mixing term in χ is on the same order of magnitude as the entropy terms, but even if the charge density were increased by a factor of 10, the electrostatic contribution to ΔG^{II} would provide only a third-order correction as long as the Debye–Hückel approximations are used.

TABLE V

The Free Energies of Phases at Equilibrium According to the Dilute Aggregate Model, $C_M = 0.3\%$, $40°$

Dilute phase, $\varphi_2^I = 0.0016$

$$\frac{\Delta G^I}{RT} = \frac{\varphi_2^I}{M} \ln \varphi_2^I + \frac{\varphi_2^I}{2M} \ln \frac{V}{v} - \frac{B_2}{(2M)^2 v} \varphi_2^I (1 - \varphi_2^I) = +1.51 \times 10^-$$

Combinatorial term	-10.4×10^{-8}
Standard state change term	$+43.7 \times 10^{-8}$
Second virial coefficient term	-1.51×10^{-4}

Concentrated phase, $\varphi_2^{II} = 0.0299$

$$\frac{\Delta G^{II}}{RT} = (1 - \varphi_2^{II}) \ln (1 - \varphi_2^{II}) + \frac{\varphi_2}{M} \ln \varphi_2^{II} + 0.6(\varphi_2^{II})(1 - \varphi_2^{II}) - Y\sigma^{3/2}(\varphi_2^{II})^{3/2}$$

$$= -1.21 \times 10^2$$

Combinatorial terms	
Solvent	-2.95×10^{-2}
Solute	-1.05×10^{-6}
Interaction terms	
Heat ($\chi = 0.6$)	$+1.74 \times 10^{-2}$
Electrostatic	$+1.71 \times 10^{-6}$

The critical conditions for phase separation are difficult to evaluate for the symmetrical aggregate model because neither chemical potential equation applies to both phases. Moreover, the experiments with symmetric gelatin mixtures have not yet been refined to provide monodisperse systems so that one phase coacervates can be obtained. Equally true is the fact that Eq. 86 and similar expressions cannot be easily used in the predictive sense because of their complexity. Nevertheless, Eq. 86 and the numerical examples based on experimental data are important in that they provide a much clearer picture of the qualitative aspects of complex coacervation and show directly those areas where further theoretical study is required.

C. A Reinterpretation of Complex Coacervation

In spite of the emphasis placed on the electrostatic factors in complex coacervation, the preceeding discussion has shown that it is possible to consider demixing in symmetric mixtures of low charge density polyions to a very good approximation without taking the electrostatic interactions into explicit account or, alternatively, by considering the interaction energy as approximately equal in both phases after demixing. These approaches both show the dominant influence of a substantial endothermic χ_{12} term in the over-all free energy expression for each phase at equilibrium, and the data indicate that the same χ_{12} is applicable in both phases.

In the temperature range studied with the gelatin–gelatin system, both gelatins are soluble over the entire concentration range examined at their respective isoelectric points. This indicates that $(\chi_{12})_{P,Q} < 0.5$. Both gelatins are polyampholytes and at their isoelectric points bear a total of nearly 200 charged groups/10^5 g. It is the solvation of these groups that keeps the gelatin in solution. The P^+Q^- aggregates, however, show effective $\chi_{12} \approx 0.6$. The model of the aggregates for which the symmetrical aggregate equations were developed is depicted in Fig. 30A where the only restriction on the distribution of charges and polymer segments is that the centers of gravity of each coincide within some small volume element. In view of the increase in χ_{12} to ≥ 0.6 in the aggregate, a more reasonable model might be that of Fig. 30B where the other extreme of complete electrostatic crosslinking is taken. The aggregate in this model has a random distribution of paired segments, or in other words, would behave like a completely flexible ladder. Hence, within the aggregate domain the "average" concentration of polymer segments would remain low, on the order of $\theta_0 = 0.01$. The charge pairing would probably be enhanced by the large numbers of positive and negative charges per molecule once the polyions are brought to overlap by the interaction of the excess net charge.

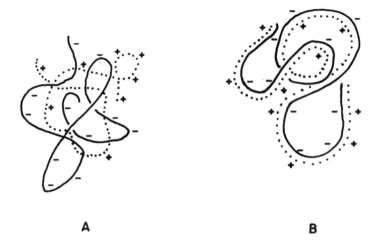

A   **B**

Fig. 30. Models for the structures within neutral aggregates. (A) Random aggregates; no correlation between charged groups but centers of gravity coincide (B) Charge pairing; correlation between segments but random distribution of paired segments.

Since the same χ_{12} applies to both phases at equilibrium, the charge pairing must persist in the concentrated phase as well as in the dilute phase. Random mixing of the neutral aggregates would permit the increase in entropy of mixing characteristic of the concentrated phase. Use of the paired aggregate in the concentrated phase would require the use of $2M$ rather than M in Eq. 40 and the inclusion of a standard state change term such as $(\phi_2/2M) \ln (V/v)$ in the free energy, Eq. 39. The numerical examples show that these modifications would make a negligible change in the previously calculated values of either $\Delta\mu_1^{II}/kT$ or $\Delta G^{II}/kT$. The point of this discussion is to stress that the demixing seen in complex coacervation is entirely analogous to the demixing of a simple polymer–solvent binary mixture as described in Section I.B. Where two distinct PQ aggregates exist, such as is presumed to be the case in the pauci-disperse gelatin mixtures, then mixing of $(PQ)_2$ and $(PQ)_3$ must occur with $\chi_{23} > 0$, and according to Eqs. 25–27 the aggregates are confined to separate phases. In contrast to Scott's assumption (see Section I.C.) that χ_{12} and $\chi_{13} = 0$, so that each solute component is miscible with solvent in all proportions, we have $\chi_{12} \approx \chi_{13} > 0.5$. Hence, in addition to the coexisting $(PQ)_2$ and $(PQ)_3$ phases, there is also a dilute phase in equilibrium with the several concentrated phases. As stated earlier, multiphase coacervate systems are the rule rather than the exception in heterodisperse mixtures. Scott's (*12*) analysis of the phase separation equilibria, however, must also be modified by the use of the proper dilute solution free energy equation, such as the virial coefficient expression, since the Flory–Huggins equation (Eq. 6) is not applicable at low φ_2.

From the point of view of the electrostatic interactions and microion salt effects, the use of the Debye–Hückel distribution of charges is clearly inappropriate, and attention should be focused on the first step of the interaction sequence, namely,

$$P^+OH^- + H^+Q^- \rightarrow (PQ)_{agg} + H_2O$$

with the assumption that all charges be paired. The correct electrostatic free energy expression for this reaction has not been determined. It seems likely that an approach similar to that taken by Rice and Harris (63) and Rice and Nagasawa (64) would be suitable. The major elements in such a treatment would be: (1) An emphasis on near-neighbor interactions by determining direct discrete site–site interaction energies, (2) the inclusion of a low dielectric constant in the interaction term to represent the limited solvation between charged groups, and (3) the summing of all interactions including the total numbers of charged groups rather than the net charges when polyampholytes are considered. The temperature dependence of the enhancement ratio ϵ was shown earlier (Section II.C.1.c) to correspond to an exothermic enthalpy of interaction, but no source of this energy was indicated. The ΔH values probably reflect the charge pairing within the aggregates in both phases, and hence would apply to the aggregate formation reaction. In the absence of interactions other than those of the fixed ionic groups, these ΔH's could be treated as the electrostatic free energy contribution and, hence, represent the point of departure for comparison of theory and experiment. The interaction energy is on the order of a few times kT, whereas the net free energy change on transfer of a polymer segment from one phase to the other is on the order of $\leq kT$. Although detailed computations taking direct interactions into account as outlined above have not yet been carried out, some qualitative predictions can be given relevant to the phase separation problem.

The reduced solvation of the ionic groups upon pairing charges appears to lead to the development of increased χ_{12}, which favors demixing. The addition of microion salts influences the demixing, reducing the concentrated phase concentration and increasing the dilute phase concentration, by decreasing χ_{12} and shifting the aggregate from the structure depicted in Fig. 30B toward that of 30A. On this basis, pair-wise aggregates do not need to be completely disrupted at very low salt concentrations but salt suppression will be evident because of the effective decrease in χ_{12} in *both* phases. Adequate experimental verification of the persistence of the aggregates throughout the salt-suppression range is not available, but attempts at the electrophoretic separation of the components of gelatin aggregates at moderate salt concentrations were not successful (Veis and Mussell, unpublished results). On the other hand, at high salt concentrations the components of gelatin aggregates were readily separated by ion exchange chromatography. Viscosity studies (*33*) indicated that the aggregates, once formed at zero salt concentration, were quite stable to either dilution or the addition of minute amounts of salt. Some time was required for reequilibration of coacervate equilibria if salt was added at low concentrations to coacervates formed in the complete absence of salts (Veis and Savner, unpublished results).

Self-suppression in heterogeneous mixtures can also be treated in electrostatic terms. In heterogeneous mixtures at $C_M \geq \theta_0$ the domains of each polymer must overlap; aggregates develop (*34*) but they may not contain two polyions of identical molecular weight and, hence, will have uncompensated charges. If the polyions are polyampholytes as in the gelatin mixtures, the additional charged groups serve to further reduce χ_{12} for the aggregate. This is a much more satisfactory interpretation than that on which the symmetrical aggregate model was originally based because the standard state entropy terms in the chemical potential expression for the dilute phase are so small. The argument presented earlier regarding the increased number of molecules of different weights in an aggregate required to maintain neutrality is still valid, however. Because of the difficulty in charge pairing, this aggregate will have the character of Fig. 30A rather than of 30B and, hence, χ_{12} will decrease below the critical value although aggregates are still formed.

IV. BIOLOGICAL IMPLICATIONS OF POLYELECTROLYTE PHASE SEPARATIONS

The development of an adequate quantitative theory of polyelectrolyte phase separation is clearly far from realization. However, even at this stage many interesting implications may be drawn from the general demixing phenomena with regard to the organization, specificity, and properties of biological systems.

Mixtures of oppositely charged polymers exhibit the following major properties based entirely upon nonspecific electrostatic interactions between the polymers:

1. *Polymer electrostatic equivalence* in concentrated phases.
2. *Selection* by molecular weight or charge density.
3. *Partitioning* into coexisting phases in heterogeneous mixtures although the chemical character of all polymers may be similar.

4. *Entropic regulation* of the concentrated phase physical properties.
5. *Enhancement or destabilization* of specific structures.

The most basic of these attributes is the requirement for very near electrical equivalence between polymer components in the concentrated phases and, hence, the maintainence of a nearly constant composition for for a concentrated phase even with extreme variation in the mixing ratio of the polymers. Oparins' (6,7) hypothesis on the use of the complex coacervation process in the prebiologic gathering together of heterogeneous randomly formed polymeric polyions from very dilute solutions is strongly supported by this requirement for equivalence which also leads directly to the molecular weight selectivity which is a prominent feature of coacervation. On the basis of both the minimization of the system electrostatic free energy in the aggregates and the concentrated phase, and the regulation of χ_{12} by charge pairing, aggregation occurs most readily from dilute solution between molecules of similar charge density, charge distribution, and molecular weight, and demixing takes place preferentially with those aggregates with the highest molecular weight [χ_{12} is a function proportional to the molecular weight (4)]. Another interesting conclusion from the charge pairing aggregate model is that in a mixture of polyions of single charge type, say A^+ and B^-, and of polyampholyte character, P^+ and Q^-, then, if the excess net charge density of each was $+\sigma$ or $-\sigma$, the most stable aggregates formed would be (PQ) and (AB). (PB) and (AQ) would have smaller values of χ_{12} and conform more of the Fig. 30A model. Hence, completely nonspecific interactions between random chain polymers can go a very long way in directing the organization of mixtures of the type of polymers commonly recognized as biopolymers.

An additional feature of the organization is the partitioning of the condensed phases, all in equilibrium with the same dilute solution and all containing the same solvent. Again, because of the fact that each paired aggregate represents a new component and that different components of varying molecular weight or charge density would not have a tendency to mix, $\chi_{23} > 0$ and the aggregates would collect into separate phases when χ_{12} for each aggregate exceeded its critical value. The coacervation system in heterogeneous polymer mixtures thus becomes the prototype of the cell in which the numerous organelles of similar character coexist in equilibrium with a common, more dilute phase. The partitioning can occur without membranes bounding the separate phases and as the result of entirely nonspecific interactions.

The concentration of the condensed or concentrated phase is obviously regulated by the distributive properties of the polymers. The higher the total charge density within an aggregate, the more concentrated the condensed phase will be. Again this is a property more dependent upon total numbers of compensated charged groups than on the net charge density. The introduction of specific structure into one or both the partners in an aggregate would tend to reduce the "saving" entropy of mixing term which favors dilution of the aggregates with solvent in the condensed phase, and the system would be pushed in the direction of coprecipitation, as seen in the polypeptide–native DNA interaction systems (Section II.D). It may be just this feature of the coacervation system, however, that makes it most applicable to questions of the behavior of protein–polysaccharide complexes in connective tissue ground substance and to the organization of the connective tissues in general. It has been established that different connective tissues within the same animal contain differing amounts of the several types of acid mucopolysaccharides and that the distribution of these vary with the age of the host. There are corresponding variations in the concentration and physical properties of these tissues.

As a corollary of the influence of specific backbone structure on the phase concentrations, one may also see the reverse effect of aggregate formation stabilizing or destabilizing a particular structure. For example, in the gelatin–gelatin mixture one may study the optical rotation of the dilute aggregate phase or the gel strength of the concentrated phase as a function of the temperature. Upon cooling a solution of aggregates formed at 80° down to some temperature below the usual gelation temperature, one observes that the specific rotation of the aggregate begins to become levorotatory at about 55–60° (Veis, unpublished results) rather than at the usual range of 25–30° as in Fig. 24. The cooperative phenomena seen in Fig. 24 is apparently the result of hydrogen bonding interactions between aggregates leading to an effective increase in their molecular weights. In contrast, the aggregates themselves at higher temperatures show a marked enhancement toward collagenfold formation, presumably because of the charge pairing interaction. This may account for the increase in C_C at $T < 55°$ noted in Fig. 21. In contrast, Felsenfeld et al. (49) demonstrated that native RNAse binds more strongly to denatured DNA than to native DNA, indicating that charge compensation is easier in the denatured DNA system. This interaction shifts the equilibrium melting temperature of DNA in the DNA–RNAse aggregate to lower temperature, destabilizing DNA. Polypeptides, which can pair charges much more effectively with DNA, stabilize the DNA in the aggregate and raise T_M by as much as 20° (50). The liquid–liquid demixing system thus is a very good model for quantitative analysis of a number of interesting biological problems, even including regulation of protein biosynthesis.

Finally, it should be noted that any type of exergonic polymer–polymer interaction leading to pairing and an increase in χ_{12} for the aggregate will lead to behavior of the complex coacervation type although the parameters regulating the interaction may not be electrostatic. For example, aqueous solutions of poly(ethylene oxide) and un-ionized poly(acrylic acid) form water-insoluble complexes by the cooperative formation of hydrogen bonds between ether oxygens and the carboxyl group hydrogens (65). In this case, the temperature plays the same role in decreasing the tendency to demixing as does the microion salt concentration in complex coacervation.

ACKNOWLEDGMENT

Most of the work from the author's laboratory described in this review has been supported by Grant GM-08114 from the National Institute of Health, U.S. Public Health Service. The author is also indebted to the John Simon Guggenheim Memorial Foundation for its support during the preparation of this review.

REFERENCES

1. A. R. Shultz and P. J. Flory, *J. Amer. Chem. Soc.,* **74**, 4760 (1952).
2. A. Dobry and F. Boyer-Kawenoki, *J. Polym. Sci.,* **2**, 90 (1947).
3. P. A. Albertsson, *Biochim. Biophys. Acta,* **27**, 378 (1958).
4. P. J. Flory, *Principles of Polymer Chemistry,* Cornell Univ. Press, Ithaca, New York, 1953, p. 554.
5. H. B. Bungenberg de Jong and H. R. Kruyt, *Kolloid-Z.,* **50**, 39 (1930).
6. A. I. Oparin, *The Origin of Life on Earth,* Oliver and Boyd, London, 1957.
7. A. I. Oparin, *Life, Its Nature, Origin and Development,* Academic, New York, 1961, pp. 39–90.
8. A. I. Oparin, K. B. Serebrovskaya, and S. A. Pantskhava, *Dokl. Akad. Nauk SSSR,* **151**, 234 (1963).
9. P. J. Flory, *J. Chem. Phys.,* **10**, 51 (1942).
10. M. L. Huggins, *J. Phys. Chem.,* **46**, 151 (1942).
11. P. J. Flory, *Principles of Polymer Chemistry,* Cornell Univ. Press, Ithaca, New York, 1953, p. 544.
12. R. L. Scott, *J. Chem. Phys.,* **17**, 279 (1949).
13. M. J. Voorn, *Rec. Trav. Chim.,* **75**, 317 (1956).
14. M. J. Voorn, *Rec. Trav. Chim.,* **75**, 1021 (1956).
15. H. G. Bungenberg de Jong, in *Colloid Science,* Vol. II (H. R. Kruyt, ed.), Elsevier, Amsterdam, 1949, pp. 232–480.
16. H. G. Bungenberg de Jong and W. A. L. Dekker, *Kolloid Chem. Beih.,* (a) **43**, 143 (1935); (b) **43**, 213 (1936).
17. W. M. Ames, *J. Sci. Food Agr.,* **3**, 454 (1952).
18. A. Veis, J. Anesey, and J. Cohen, *J. Amer. Leather Chem. Asso.,* **55**, 548 (1960).
19. J. W. Janus, A. W. Kenchington, and A. G. Ward, *Research (London),* **4**, 247 (1951).
20. A. Veis, J. Anesey, and J. Cohen, in *Recent Advances in Gelatin and Glue Research* (G. Stainsby, ed.), Pergamon, London, 1958, pp. 155–163.
21. A. Veis and C. Aranyi, *J. Phys. Chem.,* **64**, 1203 (1960).
22. W. M. Ames, *J. Sci. Food Agr.,* **3**, 579 (1952).

23. A. Veis and E. Bodor, in *Structure and Function of Connective and Skeletal Tissues* (S. Fitton-Jackson et al., eds.), Butterworth, London, 1965, pp. 228–235.
24. M. J. Voorn, *Rec. Trav. Chim.,* **75,** 925 (1956).
25. H. G. Bungenberg de Jong and R. F. Westekamps, *Biochem. Z.,* **234,** 367 (1931).
26. H. G. Bungenberg de Jong and E. G. Hoskam, *Koninkl. Ned. Akad. Wetenschap. Proc.,* **45,** 585 (1942).
27. A. S. Michaels and R. G. Miekka, *J. Phys. Chem.,* **65,** 1765 (1961).
28. A. S. Michaels, L. Mir, and N. S. Schneider, *J. Phys. Chem.,* **69,** 1447 (1965).
29. A. S. Michaels, G. L. Falkenstein, and N. S. Schneider, *J. Phys. Chem.,* **69,** 1456 (1965).
30. H. G. Bungenberg de Jong and C. von der Meer, *Koninkl. Ned. Akad. Wetenschap. Proc.,* **45,** 490 (1942).
31. J. W. Janus, in *Recent Advances in Gelatin and Glue Research* (G. Stainsby, ed.), Pergamon, London, 1958, p. 265.
32. M. J. Voorn, *Rec. Trav. Chim.,* **76,** 1021 (1956).
33. A. Veis, E. Bodor, and S. Mussell, *Biopolymers,* **5,** 37 (1967).
34. A. Veis, *J. Phys. Chem.,* **67,** 1960 (1963).
35. A. Veis, S. Mussell, and S. Savner, Unpublished Results.
36. F. W. Tiebackx, *Kolloid-Z.,* **8,** 198 (1911); **9,** 61 (1911); **21,** 102 (1922).
37. A. Veis, *J. Phys. Chem.,* **65,** 1798 (1961).
38. P. H. von Hippel and W. F. Harrington, *Biochem. Biophys. Acta,* **36,** 427 (1959).
39. M. P. Drake and A. Veis, *Biochemistry* **3,** 135 (1964).
40. E. Chargaff, M. Ziff, and D. H. Moore, *J. Biol. Chem.* **139,** 383 (1941).
41. E. Goldwasser and F. W. Putnam, *J. Phys. Colloid Chem.,* **54,** 79 (1954).
42. R. V. Rice, M. A. Stahmann, and R. A. Alberty, *J. Biol. Chem.,* **209,** 105 (1954).
43. H. Morawetz and W. L. Hughes, Jr., *J. Phys. Chem.,* **56,** 64 (1952).
44. M. Berdick and H. Morawetz, *J. Biol. Chem.,* **206,** 959 (1954).
45. R. F. Steiner, *Arch. Biochem. Biophys.,* **47,** 56 (1953).
46. J. Monod, J. P. Changeux, and F. Jacob, *J. Mol. Biol.,* **6,** 306 (1963).
47. R. C. C. Huang, J. Bonner, and K. Murray, *J. Mol. Biol.,* **8,** 54 (1964).
48. J. Coleman and H. Edelhoch, *Arch. Biochem. Biophys.,* **63,** 382 (1956).
49. G. Felsenfeld, G. Sandeen, and P. H. Von Hippel, *Proc. Nat. Acad. Sci., U.S.,* **50,** 644 (1963).
50. D. E. Olins, A. L. Olins, and P. H. von Hippel, *J. Mol. Biol.,* **24,** 157 (1967).
51. M. Leng and G. Felsenfeld, *Proc. Nat. Acad. Sci., U.S.,* **56,** 1325 (1966).
52. A. M. Liquori, L. Constantino, V. Crescenzi, V. Elia, E. Giglio, R. Puliti, M. de Santis Savino, and V. Vitagliano, *J. Mol. Biol.,* **24,** 113 (1967).
53. M. J. Voorn, *Rec. Trav. Chim.,* **75,** 405 (1956).
54. M. J. Voorn, *Rec. Trav. Chim.,* **75,** 427 (1956).
55. M. J. Voorn, *Fortschr. Hochpolymer. Forsch.,* **1,** 192 (1959).
56. J. Th. G. Overbeek and M. J. Voorn, *J. Cell. Comp. Biol.,* **49,** Suppl. 1, 7 (1957).
57. J. Th. G. Overbeek and M. J. Voorn, *J. Polym. Sci.,* **23,** 443 (1957).
58. T. A. Orofino and P. J. Flory, *J. Chem. Phys.,* **26,** 1067 (1957).
59. T. A. Orofino and P. J. Flory, *J. Phys. Chem.,* **63,** 283 (1959).
60. R. E. Gates, Doctoral Dissertation, submitted to the Graduate School, Northwestern University, June 1968.
61. F. J. Wall, *J. Chem. Phys.,* **10,** 485 (1942).

62. F. J. Wall, *J. Chem. Phys.,* 11, 527 (1942).

63. S. A. Rice and F. E. Harris, *J. Phys. Chem.,* 58, 733 (1954).

64. S. A. Rice and M. Nagasawa, in *Polyelectrolyte Solutions,* Academic, New York, 1961, p. 249.

65. K. L. Smith, A. E. Winslow, and D. E. Petersen, *Ind. Eng. Chem.,* 51, 1361 (1959).

AUTHOR INDEX

Numbers in parentheses are reference numbers and indicate that an author's work is referred to although his name is not cited in the text. Numbers in italics show the page on which the complete reference is listed.

A

Adams, A., 124, *129*
Adler, M., 69(12), 97(81), *126, 128*
Akinrimisì, E. O., 83–84(46), 108–109 (46), *127*
Akkerman, F., 168(226), *206*
Alberts, B. M., 116(105), *129*
Albertsson, P. A., 212(3), *271*
Alberty, R. A., 246(42), *272*
Alexandowicz, Z., 149(128, 129), *203*
Allewell, H., 3(5), *60*
Altermatt, H., 161, 172(196), 180(196), *205*
Ambrose, E. J., 187(283), *207*
Ames, W. M., 226(17), 227, *271*
Amos, H., 69(10), *126*
Anderson, A. J., 199(348–349), *209*
Anderson, B., 132(4), 139(59), 177 (244), *200, 206*
Anderson, D. M. W., 144(87–92), 153 (87, 91), 171(90, 91), 172(234, 235), *202, 206*
Anderson, G., 187(280), 193(302), *207*
Andreseva, A., 132(7), *200*
Anesey, J., 226(18, 20), 232(20), *271*
Anseth, A., 152(151, 152), 170(151), *204*
Antonini, E., 41(87), *62*
Antonopoulos, C. A., 199(334), *208*
Apgar, J., 121(109), *129*
Arakawa, K., 153(163), 171(163), 196, *204, 208*
Aranyi, C., 226(21), 232(21), 234(21), 243–244(21), *271*
Armstrong, J. McD., 31(72), *62*
Aschaffenburg, R., 14(38), 23(38), *61*
Aspinall, G. O., 143, 146, *202*
Astbury, W. T., 181(253), *206*
Aune, K. C., 43(91b), *62*

B

Baker, G., 197(323), *208*
Balázs, E. A., 133(15), 150(317), 164 (212), 165(214), 195(311), 196(314, 317), 198(325), *200, 205, 208*
Baldwin, R. L., 97(82), 98(83, 84), 99 (83), 100(85), 102(85), *128*
Ballantyne, M., 187(279), 191(296–297), *207*
Bamburg, J. R., 146(108), *203*
Barclay, R., 85(48), *127*
Barker, S. A., 145(100), *202*
Barlow, G. H., 153(153–154), 170(153–154), 177(232), *204, 206*
Barrett, A. J., 141(71), *202*
Barry, G. T., 145(95, 96), *202*
Basch, J. J., 31(68, 70), 32–33(68), 34 (68, 70), 41(70), *62*
Bassett, K. H., 184(266), *207*
Bayley, S. T., 182(256), 198(256), *206*
Beck, R. M., 125(114), *129*
Beers, R. F., 66–67(2), 79(28), 82(37), 94–97(71), 105–106(28, 37, 91), *126, 128*
Benesi, H. A., 164(213), *205*
Berdick, M., 246(44), *272*
Berger, A., 199(358), *209*
Bernfeld, P., 199(353), *209*
Bertin, P., 137(46), *201*
Bettelheim, F. A., 134(25), 159(188), 177, 178, 181(24, 254), 182(25, 254), 185(271), 186(175), 187, 188(254, 286), 189(286, 286a, 288–290), 190 (247, 288, 292), 191(188, 286, 289, 290, 293–295), 192(293, 295, 298), 194(308), 195(311–312a), 199(347, 360), *201, 205–207, 209*
Beychok, S., 4(17), 24, 31(59, 72), *60, 61*

SUBJECT INDEX

A

Acacia gums, structure and occurrence, 144

Acid forms, of polynucleotides, doubly-stranded helical form, 105–109

Acidic polysaccharides. (*See also* individual compounds.)
absorption spectra and dichroism, 183–185
from animal tissue, chemistry, 133–140
from bacteria, 145
behavior as gels and concentrated solutions, 188–200
complex sols, 198–200
gelation by condensation, 195–198
ionotropic gels, 193–195
swelling by vapor sorption, 188–193
cation binding, 163–166
derivatives, 146–147
diffusion-sedimentation studies, 173–175
electrophoresis, 175–178
flow birefringence, 178–180
hydrodynamic parameters, 170–171
hydrogen ion equilibria, 161–163
molecular parameters, 147–157
molecular weights and M.W. distributions, 148–155
occurrence and function, 132
physical chemistry, 131–209
from plant tissues, 140–146
size and shape of molecules, 155–157
strain birefringence, 186–188
structural properties in solid state, 180–188
structure and origin, 133–147
thermodynamics of dilute solutions, 157–166
parameters of mixing, 157–161
transport phenomena, 166–180
viscosity, 167–173
x-ray diffraction, 181–183

Adenosine-5′-monophosphate, structure, 66

Agar agar, hydrodynamic properties, 171
physical parameters, 153

Alginates, hydrodynamic parameters, 171
physical parameters, 153
structure and occurrence, 141

Alkaline forms of poly A and poly C, properties, 83–85

Amino acids, as ionizable groups in globular proteins, 5

Amylopectin sulfate, hydrodynamic properties, 171

Amylose sulfate, hydrodynamic properties, 171
physical parameters, 154

Animal polysaccharides, structure and origin, 133–140

Arabic acid hydrodynamic parameters, 171
physical parameters, 153

Arthrobacter viscosus, acidic polysaccharide from, 146

Ascophyllum nodosum, acidic polysaccharide from, 141–142

B

Bacillus subtilis, structure and occurrence, 145

Bacteria, acidic polysaccharides from, 145–146

Bases, in nucleic acids, 68–69

Black wattle gum, structure and occurrence, 145

C

Carbonic anhydrase, titration curve of, 31–32

Carboxyethyl cellulose, 147

Carboxymethyl cellulose, hydrodynamic properties, 171
physical parameters, 154